Biological Physics

Biological Physics

Raymond Kemp

STATES
ACADEMIC PRESS
www.statesacademicpress.com

Published by States Academic Press
109 South 5th Street,
Brooklyn, NY 11249, USA
www.statesacademicpress.com

Biological Physics
Raymond Kemp

International Standard Book Number: 978-1-63989-076-7 (Hardback)

Cataloging-in-Publication Data

Biological physics / Raymond Kemp.
p. cm.
Includes bibliographical references and index.
ISBN 978-1-63989-076-7
1. Biophysics. 2. Physics. 3. Biology. I. Kemp, Raymond.
QH505 .B56 2022
574.191--dc23

Contents

Preface

Biophysics is an interdisciplinary science which focuses on studying the phenomena related to various biological systems using the principles and methods of physics. It covers the entire spectrum of biological organizations ranging from biomolecules to entire ecosystems. Molecular biophysics seeks to address the questions related to molecular biology and biochemistry. It makes use of several imaging techniques such as crystallography, spectroscopy, electron microscopy and small angle scattering for visualizing biological structures. Molecular biophysics draws on the principles of thermodynamics, chemical kinetics and statistical mechanics to study complex biological events. Some major areas of research under this field are neuroscience, bioengineering, DNA structure and analysis, computer modeling and medical physics. This book is compiled in such a manner, that it will provide in-depth knowledge about the theory and practice of biological physics. Most of the topics introduced herein cover new techniques and the applications of this discipline. Through this book, we attempt to further enlighten the readers about the new concepts in this field.

Given below is the chapter wise description of the book:

Chapter 1- The branch of science which deals with the application of theories and methods of physics to study biological phenomena is known as biophysics. It overlaps with numerous fields for research such as systems biology and quantum mechanics. This resulted in the fields of systems biophysics and quantum biophysics respectively. This is an introductory chapter which will introduce briefly all the significant aspects of biophysics.

Chapter 2- Molecular biophysics is an interdisciplinary field of research that involves concepts of engineering, physics, chemistry, biology and mathematics to understand biomolecular systems. A few of the important topics studied under cellular and molecular physics are cellular function, cell division, cell cycle regulation machinery, etc. These topics are crucial for a complete understanding of the subject.

Chapter 3- The study of principles related to biological membranes including their structural and mechanical characteristics is known as membrane biophysics. It also involves the study of protein lipid interactions and transmembrane helices, nature and magnitude of cell surface charge, etc. The topics elaborated in this chapter will help in gaining a better perspective about membrane biophysics.

Chapter 4- Medical biophysics is the branch of biophysics that makes use of physics to describe biological processes for medical applications. Some of the significant topics studied under medical biophysics are electromyography, ultrasonography, blood flow measurement, cardiac pacemakers, flow cytometry, etc. This chapter discusses in detail these significant aspects related to medical biophysics.

Chapter 5- The various topics studied in association to biophysical structure and phenomena are enzyme kinetics, molecular motor, myoglobin, hemoglobin, pancreatic ribonuclease, antibody, collagen and transfer RNA. In order to completely understand biophysical structure and phenomena it is necessary to understand the processes related to it. The following chapter elucidates these varied processes and mechanisms associated with this area of study.

At the end, I would like to thank all those who dedicated their time and efforts for the successful completion of this book. I also wish to convey my gratitude towards my friends and family who supported me at every step.

Raymond Kemp

1

Biophysics: An Introduction

The branch of science which deals with the application of theories and methods of physics to study biological phenomena is known as biophysics. It overlaps with numerous fields for research such as systems biology and quantum mechanics. This resulted in the fields of systems biophysics and quantum biophysics respectively. This is an introductory chapter which will introduce briefly all the significant aspects of biophysics.

Biophysics is a branch of science that uses the methods of physics to study biological processes. Physics uses mathematical laws to explain the natural world, and it can be applied to biological organisms and systems to gain insight into their workings. Research in biophysics has helped prevent and treat disease, advance drug development, and create technology to allow humans to live more sustainably and protect the changing environment.

Areas of Biophysics

Biophysics is incorporated into many diverse areas of biology. Some research topics in biophysics or involving biophysics include:

- Membrane biophysics: The study of the structure and function of cell membranes, including the ion channels, proteins, and receptors embedded within them.
- Computational/theoretical biophysics: Using mathematical modeling to study biological systems.
- Protein engineering: Creating and modifying proteins to advance synthetic biology. Often used to advance human health in the form of new disease treatments.
- Molecular structures: Biophysics studies the molecular structures of biological molecules including proteins, nucleic acids, and lipids.
- Mechanisms: Using physical mechanisms to explain the occurrence of biological processes. Some physical mechanisms include energy transduction in membranes, protein folding and structure leading to specific functions, cell movement, and the electrical behavior of cells.

Quantum Biophysics

Quantum biophysics is a multidisciplinary area of molecular biophysics and quantum biology concerned with quantum treatments and theories of biomolecules and biophysical systems in general at the quantum level. Quantum biophysics is considered as a subject where biological systems are understood through the application of physical principles and theories. Quantum mechanics has long been known to play a crucial role in the world of chemistry and in turn, chemistry has been able to set the structured, predictable rules for biological systems.

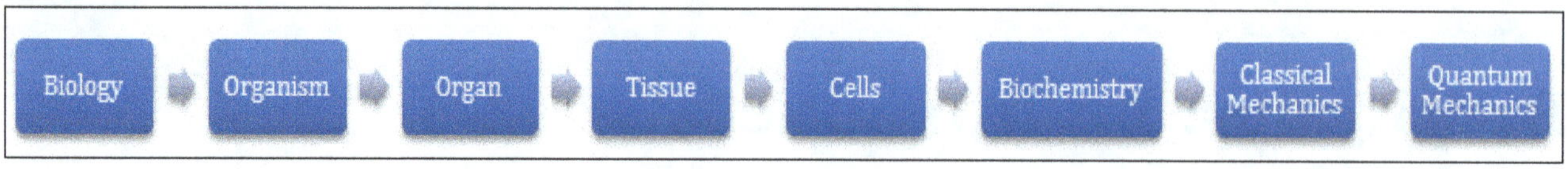

As lower levels are reached atoms and subatomic particles are observed. Their relationships and behaviors are explained with quantum mechanics and lead to the many reactions seen in biological systems.

The effects of quantum mechanics can of course be observed, but usually only under extremely rigid conditions such as temperatures near absolute zero and under the influence of vacuums.

Quantum Tunnelling

The Sun

The energy provided from the Sun drives life on Earth, from metabolic reactions to reproducing behaviors. In order to supply this energy though, the Sun performs special types of reactions – fusion reactions. The sample hydrogen particles in the Sun fuse together to form helium and other heavier elements, releasing huge quantities of energy as a result. Essentially, the Sun is converting its mass into energy.

Since the 1920's, we have been researching and attempting to replicate the process of nuclear fusion, in order to harness the large amounts of energy for a variety of purposes. However, one of the main limitations on researching this nuclear process on Earth is achieving the required temperature needed for fusion to occur. The core of the Sun is estimated to be 15.6 million Kelvin and therein lays the problem.

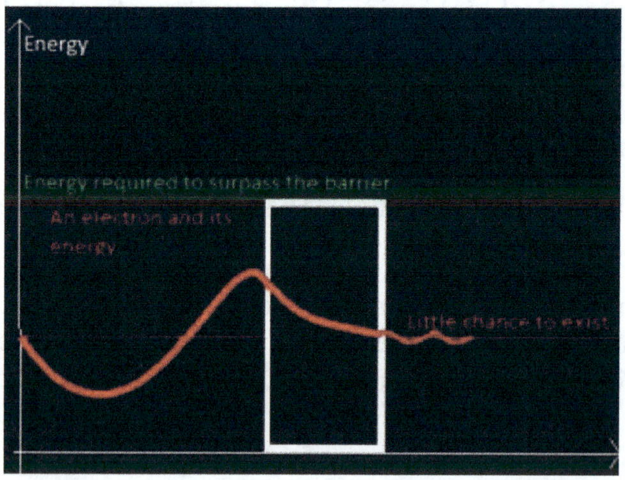

An electron does not have the required energy to get over the barrier to start a reaction. There is a small probability that it can tunnel through the barrier though and the reaction takes place as a result. This is how nuclear fusion works in the Sun, driving life on Earth.

As the hydrogen particles increase in temperature, there is a small chance that the hydrogen atoms will fuse together, even if the correct temperature has not been reached yet. This effect was mathematically introduced in 1928 by Hufauer. Balantekin and Takigawa demonstrated that rather than waiting for the required temperature needed to overcome the energy barrier that starts the fusion reaction, the electron in the hydrogen atom 'tunnels' through the barrier instead.

The significance of this process is twofold. Firstly, our main source of heat and catalyst of many biological reactions relies on quantum mechanics to provide the sheer amount of energy required to drive biology processes on Earth. And secondly, if a structure as important and relatively simple as the Sun can use quantum forces, surely the incredibly complex and high number of processes that occur in biological systems, have undiscovered quantum connections.

Enzyme Catalysis

The fastest studied reactions in the history of science include those in biological systems that are caused by proteins called enzymes. An appreciation of the sheer speed and rate of these reactions is yet to be completely realised. In the absence of enzymes, these same reactions are among the slowest that have ever been measured, some with half-times approaching the age of the Earth. After a substrate is bound at an enzyme's active site (forming an enzyme-substrate complex), its half-life is usually a small fraction of a second.

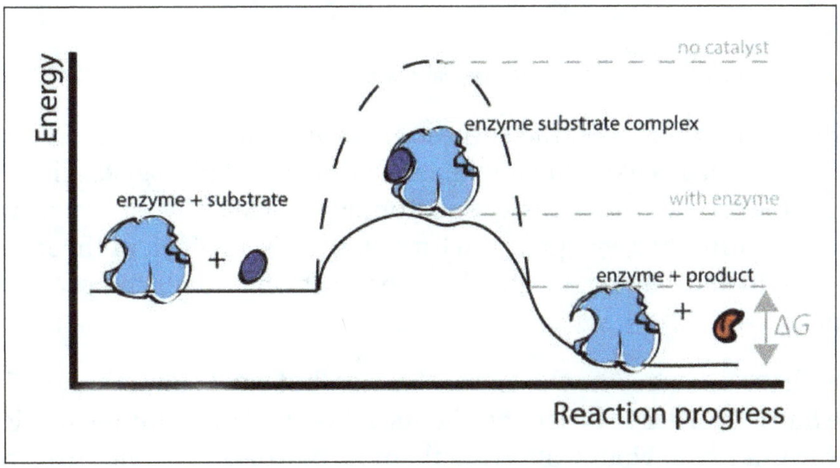

Lowering of reaction's activation energy caused by an enzyme-substrate complex. Also simulating a visual of the effects of quantum tunnelling on enzyme catalysis. Back in 60s, it was discovered that quantum tunnelling also takes place inside living cells. Amphibians, such as frogs, use quantum tunnelling to undergo the biological process of metamorphosis – where the tadpole breaks down and reassembles as a frog. The speed at which this process occurs cannot be adequately explained by traditional methods of thinking. The tail is remarkably broken down into the basic units of life, and is redistributed in the newly forming frog as proteins, such as enzymes and collagen (a protein that supports the muscles, tendons and ligaments within living organisms and has elastic properties).

The strong bonds between proteins are very difficult to separate and should last much longer than the tadpoles lifespan. Enzymes are responsible for breaking these bonds at an accelerated rate. The spontaneous rates of biological reactions examined in this way, span a range of more than sixteen orders of magnitude, with half-lives ranging from five seconds, for the hydration of CO_2 to 1.1 billion years, for the decarboxylation of amino acids.

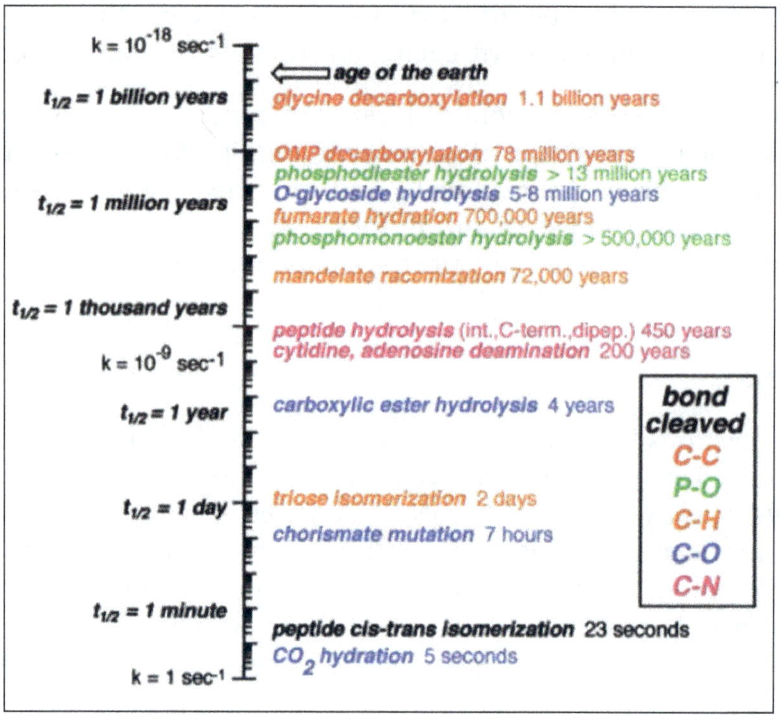

Natural half-times of some biological reactions in neutral solution at 25°C.

This incredible acceleration has been the subject of discussion and debate for nearly thirty years now and a leading theory has peaked the interests of physics and biologists alike. A group of scientists found that enzymes are able to transfer subatomic particles, like electrons and protons, from one part of a molecule to another, via quantum tunnelling. It's an efficient theory and explains how the proton tunnelling to another location will cause the strong bonds to become weak bonds in a fraction of a second.

The protons do not have to overcome the energy barrier, but can tunnel through it (similar to tunnelling within the Sun), in order to break the chemical bonds. As subatomic particles are not solid, that can exhibit wave-like behaviours, allowing them to approach a barrier from one side and they have a certain probability of reappearing instantaneously on the other.

(A) (B) (C)

In figure, (A) shows a simple but effective square knot at the top and the bottom shows a similar tie but will never join the two pieces of rope. This demonstrates the quantum effect of tunnelling in where, in this case, there are two possible outcomes–a strong bond and a weak bond. (B) shows a digitally altered image where one of the ends of rope (the proton) on top is 'tunnelling' to a new position. (C) Illustrates the new position and now has effectively caused the bond to break. This could be demonstrated by a person pulling at any end on any rope in image (C). It would instantly come undone. In metamorphosis, the collagen proteins in the tadpole tail are quickly broken down and other enzymes are then able to redistribute the particles to form the frog.

Contrasting such statements, with near overwhelming evidence, it is theorized that enzymes can use their dynamic behaviour to force protons through the energy barrier, a characteristic unique to these powerhouse proteins. Evidence is even beginning to emerge that such behaviours are selected by evolution to enhance enzyme catalysis. Future research will provide intellectually provocative models of enzyme catalysis and suggest new avenues for experimental and computational investigations.

Quantum Entanglement

Migration by Magnetoreception

The European robin (Erithacus rubecula) migrates from Scandinavia down to the Mediterranean, every autumn. Whilst migratory patterns with birds caused by seasonal changes is not a new phenomenon, what is special and worth investigating, is how they manage to travel to the exact same geographical areas each year. These areas are the same locations that their predecessors emigrated to and, hence, human curiosity has long wondered, how do organisms know exactly where they are going, particularly if they have never been there before? How can birds navigate with such accuracy? Is it a magnetic particle within them that responds to Earth's magnetic field"? Is it a behavioural trait passed from one generation to the next? Or is it something smaller than a gene and much more complicated?

The European robin, and like a lot of other marine animals and even insects, navigate by sensing the Earth's magnetic field.

Sketch of the experimental equipment and set-up in the planetarium.

These hypotheses were experimentally tested in a repurposed planetarium using specialised Emlen funnels which are round cages in the shape of funnels lined with scratch sensitive paper and an inkpad on the bottom.

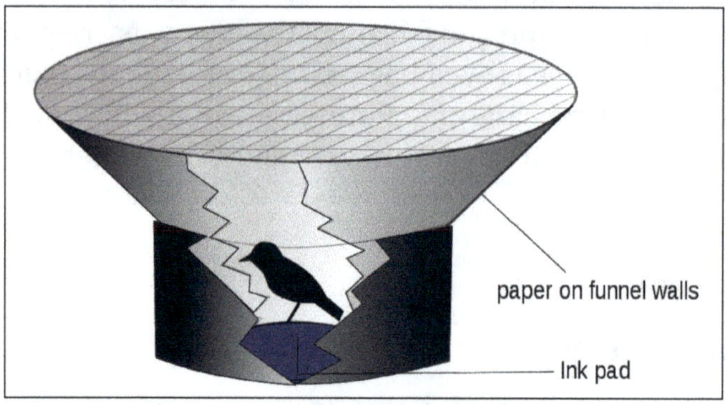

An Emlen funnel, used to document migration orientation in bird.

At night, with the artificial magnetic field inactive, the caged migrant birds became restless and jumped more often in the direction where they intend to migrate, making ink prints on the scratch sensitive paper. However, what was even more astounding was that if the wavelength of light was changed, the behaviour of the robins changed and instead of moving in the direction of expected migration, their movements became random and they jumped in a variety of directions. It seemed that monochromatic blue light (the predominant wavelength of light at night) enhanced the robin's orienteering ability.

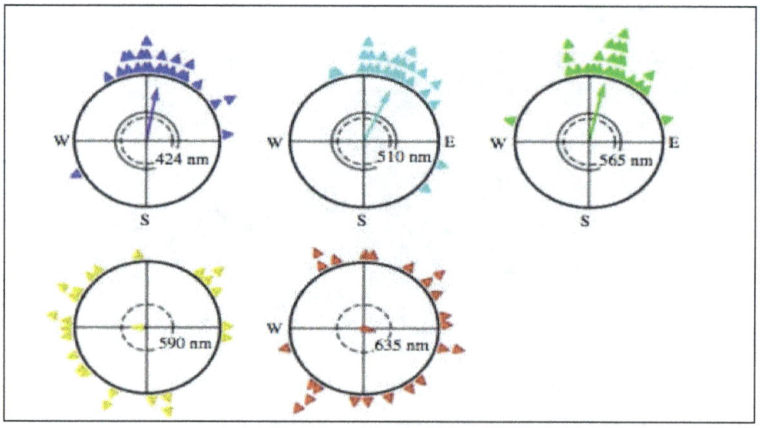

Experiments show birds exposed to monochromatic light exhibit differing orientating behaviour – orientation is better in 'blue' light.

The caged robins also changed their heading accordingly when the magnetic field was artificially altered using strong magnets, with variable directions being applied. This meant that when the magnetic field was active in a particular direction, the robin moved in the same direction. If living systems are sensitive to the magnetic vector potential component of the geomagnetic field, which is directed East-West, this mean that they must have both X and Y coordinates available. This posed an interesting dilemma, was the underlying cause of this behaviour due to wavelengths of light or a varying magnetic field, like Earth's?

It was found that photons excited a particular protein molecule, cryptochrome, in the retinal cells of the right eye and causing an electron from a particular particle to move to another. This means both molecules are now charged, that is to say one molecule is positive and one is negative. The incoherent part of this anomaly is that these molecules are both positive and negative at the same time. They exhibit behaviours of both charges until one is measured, and, hence, the other particle when measured will be the opposite charge, due to the Pauli Exclusion Principle. These particles are now in a state of quantum entanglement where they are correlated with each other, but they lose their individuality.

The infamous example of Schrodinger's Cat helps explain this unusual concept. If you were to place a cat into a box containing a randomly-released lethal gas and sealed the box up, the cat itself could be thought of as both dead and alive, by an external observer. The charged particles, exhibit such behaviours and hence can be thought of in the same way as Schrodinger's Cat.

Entangled particles are extremely sensitive to magnetic fields. Their interactions, caused by the angle of inclination of the magnetic field, trigger action potentials and the subsequent chemical reactions act as sensory chemicals which tell the robin which way to go. The entangled pair induces different chemical reactions based on their relationship and quantum state (orientation of spin). The tiny variations in Earth's magnetic field can alter the state of these entangled electrons and hence affect the robin's ability to sense its orientation. Even Einstein hated this idea: he called it "spooky action at a distance" as entanglement implied that the relationships between particles occurred faster than the speed of light, impossibility in Einstein's theory of relatively.

The leather hood used to cover the right eye of a European robin – the source of cryptochrome and the inclination compass used for migratory orientation.

The implications of research could be ground breaking. Many organisms use cryptochrome as a photoreceptor so this quantum-assisted magnetic sensing, could be widespread. Not only birds, but also some insects and even plants show physiological responses to magnetic fields. An example includes the growth-inhibiting influence of blue light on the flowering plant Arabidopsis thaliana and it is moderated by magnetic fields.

Quantum Coherence

Photosynthesis

In the cases of the avian compass, the primary evidence for quantum processes is observation of animal behaviour, combined with a lack of plausible classical models. However, by contrast, evidence for quantum influences in the biological process of photosynthesis is direct, measurable and overwhelming.

Quantum coherence is the idea of quantum entities multitasking. It can be represented by the quantum skier from figure below where an object behaves like a wave so it doesn't just move in one direction or the other, but can follow multiple pathways at the same time.

The quantum skier counter intuitively exhibits a behaviour of being able to follow several paths at once and 'finds' the most effcient one to perform its designated process.

A light harvesting, photosynthetic system uses 'antenna' molecules called chlorophyll inside chloroplasts, to capture photons, the quantum of light (sunlight energy). This capturing of photons knocks out an electron from the chlorophyll molecule and creating a quantum particle called an exciton.

The exciton travels through surrounding chlorophyll molecules, until it reaches the reaction centre and ultimately results in the production of Adenosine trisphosphate (ATP), the 'molecule of life' via aerobic cellular respiration.

For maximum efficiency, all photons absorbed have to make it to the reaction centre, but there are so many random paths and the photon does not 'know' the most efficient route. If the exciton doesn't reach the reaction centre fast enough, the energy will dissipate into the system. So how then can the process of light-harvesting, by nearly 100% efficient.

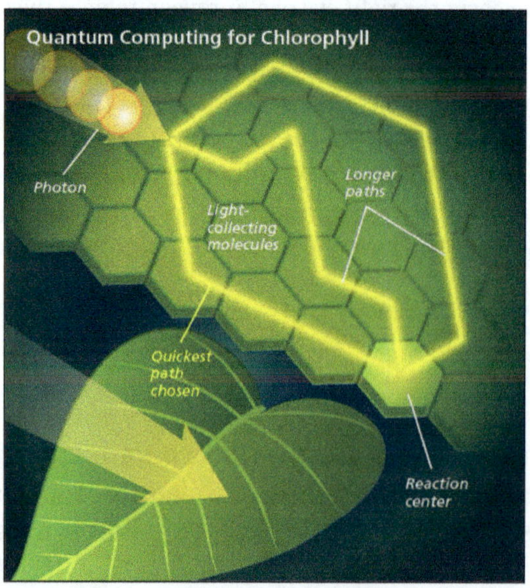

The coherence of quantum particles allows the wave-like properties to be examined and the most efficient path of the photon energy (via the exciton) will be the result.

The evidence from multiple, independent studies, is that rather than taking one path, quantum coherence is used to travel all paths simultaneously. Therefore, there is plenty of evidence that can demonstrate that the interplay between quantum coherence and the performance of biomolecular complexes. This can be intuitively explained by the fact that quantum coherence allows the system to explore several network paths in parallel to achieve the desired task to find the optimal solution and leave out inefficient slower pathways.

Quantum Biology

The Theory of Evolution – Quantum Edition

DNA is comprised of complementary nitrogenous bases (adenine and thymine, cytosine and guanine) and during the cell cycle, the majority of time is spent in a stage called interphase. This is where the cell performs its normal functions including the process of DNA replication. Essentially, this is how one strand of DNA is replicated to form 2 identical strands for the purpose of cell division (mitosis or meiosis).

The two strands of DNA found in a double-helix structure, are held together by these nitrogenous bases, like a twisted ladder. These bases are bound together by relatively weak hydrogen bonds (protons). The process of DNA replication is always proofed and checked for errors by an enzyme named DNA polymerase which can repair most mutations, however polymerase still has its limitations. Sometimes in the process, mistakes can occur and these mistakes are called genetic mutations.

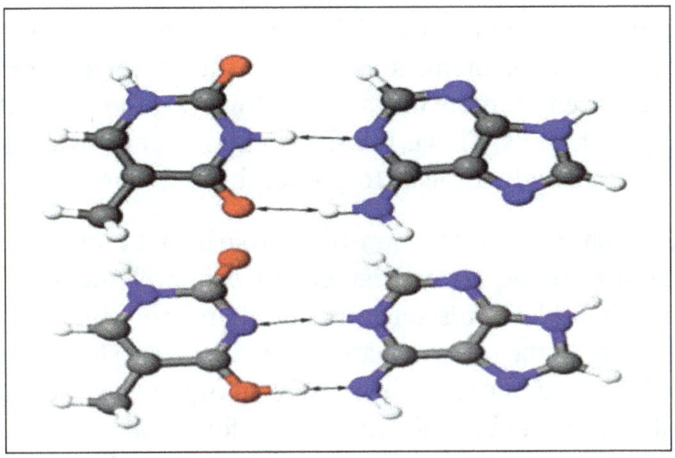

The white atoms are hydrogen and protons from the hydrogen atoms can "jump" across to the other side via quantum tunnelling. If the strands of DNA are separated by DNA replication or protein synthesis, while the hydrogen has jumped across to the opposite side, these strands may not replicate or transcribe correctly–this is a mutation.

This movement can be explained by the process of quantum tunnelling. When replication is initiated by an enzyme called DNA helicase, which splits hydrogen bonds. The two white balls in the middle of figure above are protons, one proton prefers to sit on one strand; the other, on the other strand.

They have ability to tunnel through the bonds to the other side. If the two strands of DNA then separate for replication to occur, and the two protons are in the wrong positions, this could ultimately lead to a mistake known as a mutation.

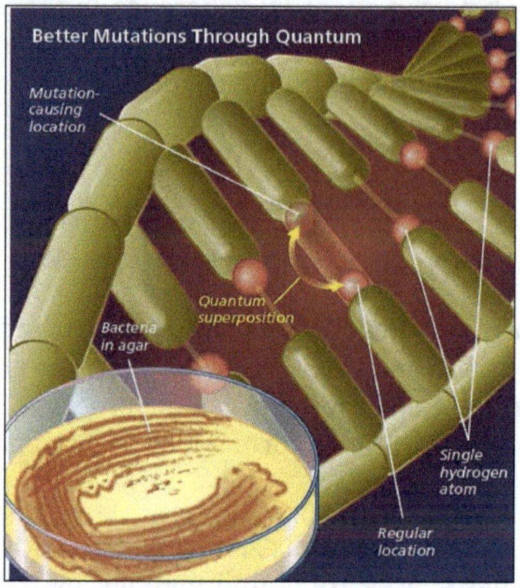

DNA mutates when this hydrogen atom tunnels, quantum-mechanically, to the "wrong" half of its rung. The property of coherence, before it is observed, the atom will simultaneously exist in both a mutated and non-mutated state-that is, it would sit on both sides of the rung at the same time.

This concept has been examined using bacteria. A population of bacteria (E. coli) with hydrogen nuclei was cultured and the mutations in the produced cells were quantified. At the same time, a separate population was cultured, but these had deuterium nuclei (deuterons), a heavier isotope of hydrogen. The results of these experiments showed that the protons within hydrogen nuclei were able to tunnel easily and thus led to many mutations. However, the population containing deuterons did not produce many mutations. This is because deuterons are twice as heavy as the hydrogen nuclei and therefore are increasingly difficult to tunnel through energy barriers.

The concept of quantum mechanics playing a non-trivial role in DNA mutations is still heavily debated today, predominated because the concentrated deuterium oxide will not take a frequency imprint until about 4% of hydrogen oxide is added. This suggests that there are 'windows' of opportunity in which biological systems are sensitive to the electro quantum environment, leading to a possible eventuality of changes in the genetic information carried by individual organisms. Current research certainly allows for quantum tunnelling to be a plausible mechanism for variations to occur and hence have the opportunity to be naturally selected to survive in certain conditions. It's even speculated that through the process of quantum coherence, living cells act as biological quantum computers and are able to explore multiple possible mutational states simultaneously, and select those states that provide the greatest advantage – the theory of Evolution redefined as a construct of quantum mechanics.

System Biophysics

Systems biophysics, in contrast, emphasizes the focus on how the system properties emerge from the relations between constituent elements. These types of approaches are needed, for instance, to study how mutations affect the molecular properties of the cellular components; how the mutated components affect different signaling pathways; and how these modified pathways confer cell-growth advantages during tumor progression and metastasis. The reductionist movement of twentieth century biological science successfully used the tools of biochemistry, molecular biology, and structural biology to provide us with an increasingly detailed parts list of living systems. As the troves of molecular data grew, the advent of bioinformatics brought to bear information technologies that allowed biological scientists to annotate, query, search, and integrate these data with relative ease. This gave birth to systems biology.

Systems biology is concerned with the study of biological functions and mechanisms, underpinning inter- and intra-cellular dynamical networks, by means of signal- and system-oriented approaches. Systems Biology's a well-established area of human activity that can be traced back at least to the writings of Aristotle around 350 B.C. The term "system" signifies here an entity that represents wholeness, which can be superficially divided (fractioned) into components (parts) but whose pivotal properties cannot be fully explained solely from the knowledge of the parts. The paradigms of life sciences were changing with time but the core fundamental concepts of systems biology are apparently resistant to these changes and remain pretty much the same today as they were in time of Aristotle.

To understand biology at the system level, we must examine the structure and dynamics of cellular and organismal function, rather than the characteristics of isolated parts of a cell or organism. Properties of systems, such as robustness, emerge as central issues, and understanding these properties may have an impact on the future of medicine. However, many breakthroughs in experimental devices, advanced software, and analytical methods are required before the achievements of systems biology can live up to their much-touted potential.

Human Systems Conceptualized to Work as Machines

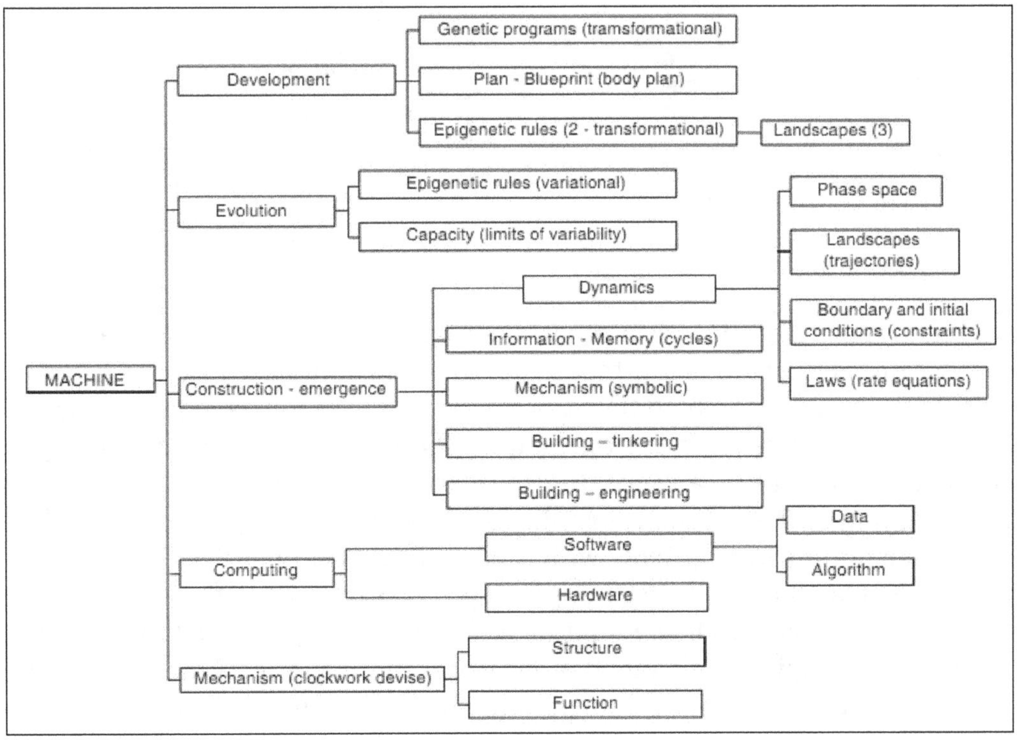

- Systems Biology is defined by assays, data types, global assays and the types of data integrations.
- Hypothesis driven Global Quantitative Integrative Iterative Dynamic Multiscale Cross-disciplinary.

Levels of Biological Complexity

- DNA,
- RNA,
- Proteins,
- Protein-protein & protein DNA Interactions,
- Pathways – Networks,
- Cells & tissue,
- Organs & Systems,
- Organisms,
- Population,
- Ecology.

Computational and Mathematical Approaches and Tools

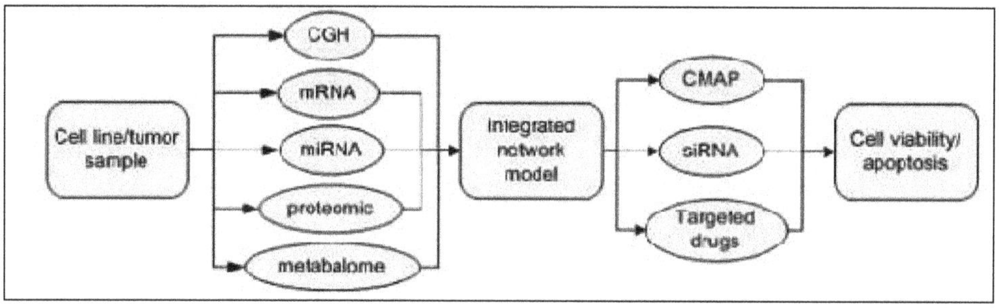

Our increased awareness that biological process is regulated by intracellular networks there has been recent focus on methods to target the networks with combinations of drugs or genetic perturbations rather than single points in the network. Targeted perturbations using small molecule inhibitors have very useful in revealing the complexity of circuits in cancer cells by uncovering regulatory feedback loops which can serve to bypass the perturbations. Perturbations of genomic expression levels using siRNA, shRNA, or miRNA show that the signaling networks are highly dynamic and capable of rewiring circuits within the cancer cells that then establish new context specific interactions that allows cells including tumor cells to persist in the presence of the perturbation. These adaptive responses likely limit the efficacy of many targeted therapeutics in use or in clinical trials. The emerging challenge that limits the significance of these efforts to functionally characterize cancer cells is how to integrate the large amount of divergent data sources into executable, predictive, system's level models that can be used to positively impact cancer treatment by determining fragility points in the networks that can be targeted for personalized cancer therapy.

Systems Biology Approach to Understand Cancer

Cancer is now appreciated as not only a highly heterogeneous pathology with respect to cell type and tissue origin but also as a disease involving dysregulation of multiple pathways governing fundamental cell processes such as death, proliferation, differentiation and migration. Thus, the activities of molecular networks that execute metabolic or cytoskeletal processes, or regulate these by signal transduction, are altered in a complex manner by diverse genetic mutations in concert with the environmental context. A major challenge therefore is how to develop actionable understanding of this multivariate dysregulation, with respect both to how it arises from diverse genetic mutations and to how it may be ameliorated by prospective treatments. While high-throughput experimental platform technologies ranging from genomic sequencing to transcriptomic, proteomic and metabolomic profiling are now commonly used for molecular level characterization of tumor cells and surrounding tissues, the resulting data sets defy straightforward intuitive interpretation with respect to potential therapeutic targets or the effects of perturbation.

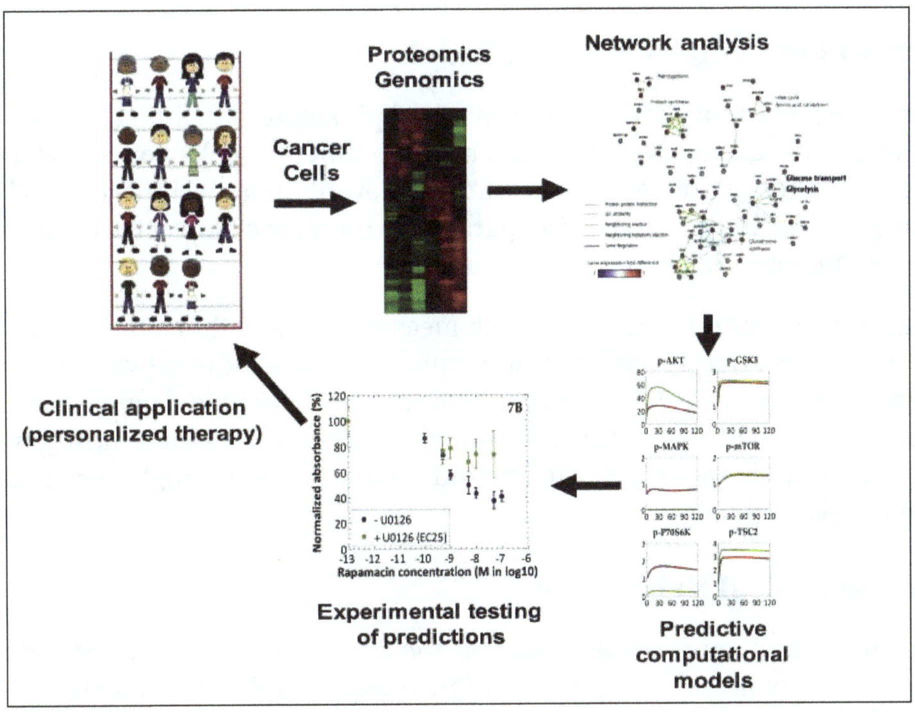

Applying Experimental High-Throughput and Whole-Genome Techniques

Molecular Modeling in Systems Biology

An important aspect of systems biology is the concept of modelling the dynamics of biochemical networks where molecules are the nodes and the molecular interactions are the edges. Due to the size and complexity of these networks, intuition alone is not sufficient to fully grasp their dynamical behaviour. Instead an explicit mathematical description of the network and its interaction dynamics is used, which allows for testing and predicting the behaviour in computer simulations. Be aware of methods for estimating model parameters and tools for model analysis of properties such as robustness.

Understand the concept of diffusion and why it can be important when creating models in systems biology. Be introduced to reaction-diffusion models Define the molecular players and interactions. It is a slow and hard process to do experimental work for elucidation of which molecules are involved in different biological processes, and how these interact. The genome projects, where the complete genomes of different species are sequenced, have increased the knowledge.

Modeling in Systems Biology

Molecules and Interactions in a Mathematical Model

To be able to do a quantitative model of a biochemical network, a system has to be defined where molecular concentrations are the variables and their interactions are described explicitly by mathematical functions. These functions depend on the type of interactions that are described, and a main goal in these lectures is to be familiar with common 'translations' for reactions, transcription, etc. into equations, with the ultimate goal of a complete quantitative model for the network.

Estimate Parameter Values for the Model

The mathematical description includes a number of parameters defining for example reaction rates. Different values of these parameters can result in different behaviours of the model. Hence it is crucial to estimate the parameter values that are relevant for a specific biological network. One possibility is to experimentally measure the parameter for a specific reaction, which will result in an optimal single estimated value for the parameter.

A potential drawback is that it is hard to do such measurements within a biological organism, and if it is measured elsewhere that specific condition might lead to a different value compared to within the organism. Another approach is reverse engineering, where model parameters are estimated by fitting model output to available experimental data. This will exclude most parameter values but still it is possible that this approach will find different values for a single parameter that equally well describe the biological behaviour.

Analyse the Dynamical Behaviour

A final step in a modelling approach is to analyse the behaviour of the defined model. Many molecular networks and modules show very high robustness. This should then also be accounted for in the model and can be tested by a sensitivity analysis, where the change of behaviour is tested when parameters are perturbed. Also, the model can be tested for perturbations where molecules or interactions are removed from the system, which then can be compared with knockout experiments, or provide biological predictions from the model. The analysis can provide feedback into the previous three steps improving the description and knowledge of the biological system.

Systems biology involves the quantitative and simultaneous integration of different and multiple biological components and their relationships with one another. For example, the components may be proteins, while their relationships may be described by signal transduction pathways.

Unlike systems biology, molecular modeling focuses on a single complex between biomolecules and computes the interactions that exist in the complex. Although the two fields appear dissimilar, they are both quantitative in nature and involve many components and relationships. In the case of molecular modeling, the components are the atoms and their partial charges, and their relationships are the different interactions between them. Therefore, it's no surprise that some molecular modeling methods are now being applied to systems biology. Moreover, there has been recent success in combining these two fields to rationally design effective therapeutics.

Applications of Systems Biology

Systems biology revolves around the principle that the phenotype of any individual living organism is a reflection of the simultaneous multitude of molecular interactions from various levels occurring at any one time. Consequently, against the standard concept of reductionist approach where dysregulations in isolated molecular components are studied, data from dysregulations of multiple key molecular players from varying cellular levels of activity are pooled and studied in their entirety, for the purpose of identifying distinct changes in the pattern of intermolecular relationships, vis-a-vis the organism's investigated phenotypes. The methods applied as the principal research tools vary, depending on the nature of the molecular level being investigated and also on the volumes of data generated; therefore, nowadays most self-sufficient systems biology research groups are composed of research scientists with a discernable knowledge of experimental investigation for most molecular level research and/or are unique experts in their own specific research field.

Consequently, systems biology is very much an interdisciplinary field of research, requiring the technology platforms and research expertise of individuals from a spectrum of scientific research niche. However, the measurement of all molecular parts of an organ or even biomedical pathway is far from routinely achievable, and great efforts to improve sensitivity of analysis and to make the output data possess a quantitative significance are starting to improve through implementation of field standards. Given current constraints, Boolean approaches are assisting with production of 1st generation systems biology models. A main difference between systems biology and systems medicine is that the former assumes the data to be correct and useable as often wet-lab data generation expertise is not the main goal but is assumed to be correct and useable. Systems medicine (sometimes referred to as systems healthcare) promises to lead with clinical and molecular know-how to produce exquisite datasets that are employed to generate pathway models and treatment and will hopefully directly contribute to stratified medicine enroute to personalized healthcare.

The essential facets of systems medicine should ideally be established in order to provide proper support for the effective and rapid implementation of any novel research methodologies aimed at reaching the intended outcome for systems medicine-based projects. Systems Medicine is more of an integrative interdisciplinary approach networking and consequent data handling and flow of information that are vital components for enabling successful systems medicine approaches.

Another important component for systems medicine involves the employment of computation of computational and modeling sciences. Such expertise is a prerequisite for the effective handling of "big" datasets and also for the interpretation of wet laboratory data in terms of the development of complex interrelationships between varying key molecular players.

Application in Neurological Sciences

Recent advances in molecular biology, genetics, and imaging have demonstrated important insights about the nature of neurological diseases. However, a comprehensive understanding of their pathogenesis is still lacking. Although reductionism has been successful in enumerating and characterizing the components of most living organisms, it has failed to generate knowledge on how these components interact in complex arrangements to allow and sustain two of the most fundamental properties of the organism as a whole: its fitness, also termed its robustness, and its capacity to evolve.

Systems biology complements the classic reductionist approaches in the biomedical sciences by enabling integration of available molecular, physiological, and clinical information in the context of a quantitative framework typically used by engineers. Systems biology employs tools developed in physics and mathematics such as nonlinear dynamics, control theory, and modeling of dynamic systems. The main goal of a systems approach to biology is to solve questions related to the complexity of living systems such as the brain, which cannot be reconciled solely with the currently available tools of molecular biology and genomics. As an example of the utility of this systems biological approach, network-based analyses of genes involved in hereditary ataxias have demonstrated a set of pathways related to RNA splicing, a novel pathogenic mechanism for these diseases. Network-based analysis is also challenging the current nosology of neurological diseases. This new knowledge will contribute to the development of patient-specific therapeutic approaches, bringing the paradigm of personalized medicine one step closer to reality.

Application in Cancer Biology

Cancer is a complex disease system involving multiple molecular, genetic, and cellular events. From its early initiation through progression and metastasis, cancer can adapt and evolve as a result of both internal and external signals. These properties make cancer difficult to predict, prevent, and treat. There has been significant progress in characterizing the genetics of cancer, as well as the downstream effects on the molecular and cellular pathways that are critical for the initiation and progression of cancer. However, studies focused on the role of one gene have highlighted the need to understand cancers as integrated systems of genes, networks, and intercellular interactions.

Cancer systems biology is uniquely poised to address the complexity associated with cancer through its unique integration of experimental biology and computational and mathematical analysis. Instead of viewing cancer through the lens of a single mutation or alteration, the goal of cancer systems biology is to provide a bird's eye view of the changing cancer ecosystem, thus allowing cancer biologists and oncologists to understand and predict how one alteration effects an entire tumor system.

Application in Drug Designing

One of the main research challenges in which systems medicine perspectives can make a major difference is in the prediction of drug adverse effects during the early phases of the drug development process. Pharmacogenetics research for the purpose of drug development has, in the past, focused almost entirely on the effect of variations in individual genes for causing a specific adverse effect. However, such adverse effects are most possibly due to multifactorial influences and as such a systems-based investigation can be far more effective in rooting out and/or predicting harmful adverse effects prior to any novel lead molecule progressing any further within the drug development pipeline. This study demonstrated that apart from playing an important part in predicting drug adverse effects, systems medicine methodologies can also be employed for the prediction of ideal drug combination therapies to be adopted for specific disease conditions.

Another area of drug development in which systems approaches are of significance is in the prediction of drug/target interactions. The study utilized the Connectivity Map (CMap) to select candidate hERG inhibitors with similar gene signature expression profile induction, together with analytical methodologies from databases of experimental datasets for annotated hERG inhibitor activities.

Other systems-led methodologies can also be implemented for the purpose of drug/target interaction prediction studies, particularly for large and heterogeneous molecular interaction networks. Such methods include probabilistic matrix factorization and Network-based Random Walk with Restart on the Heterogeneous network (NRWRH). Similarly, the importance of network pharmacology methods for the identification of novel drug/target networks for individual and/or multiple disease conditions is becoming ever more important due to the recent trend in the application of polypharmacology avenues for maximizing drug development efforts through enhanced treatment successes. This approach is of crucial value particularly for complex and multifactorial clinical conditions such as cancer pathways, bearing a wide spectrum of druggable targets.

Impact of Cancer Systems Biology

There has been an explosion in the quantity of available experimental data from high-throughput technologies, such as genomic sequencing, transcriptomics, metabolomics, and proteomics. Additionally, targeted experiments on a smaller scale have provided important information about complex interactions within and between cells. Systems analyses and predictive modeling are necessary to integrate across these datasets that span different length and time scales to convert them into knowledge. Systems Biology brought together cancer biologists and oncologists with computer scientists, mathematicians, and engineers to tackle important cancer questions including the identification of tumor master transcriptional regulators, the role of microRNAs in cancer progression and metastasis, the development of drug resistance in heterogeneous tumor contexts, and optimization of combination therapies for cancer treatment. It is envisioned that the successes of cancer systems biology will encourage investigators to bring these tools to bear on emerging and difficult cancer questions that will require a systems approach to fully comprehend.

Complex phenotypes emerge from the interactions of thousands of bio-molecules that are organized into complexes and interacting with one another as one functional module which in turn, emerges as functional networks in health and disease. Omics approaches collect data on changes for all genes or other macromolecules in the cells and statistical analysis attempts to uncover the functional modules that perform the functions that characterize higher levels of biological organization. Systems biology attempts to analyze the study of individual genes/proteins and to integrate them into one functional unit. Cancer cells exhibit an altered/defective genetic and epigenetic landscape formed by altered complexes and network modules arising in different parts of tumor tissues that sustain autonomous cell behavior ultimately promoting tumor growth. Understanding of tumor biology must address not only molecular but also, and more importantly, tumor cell heterogeneity, by considering cancer tissue genetic and epigenetic networks.

Brain Cancer

The most common types of brain cancers are astrocytomas, which are malignant tumors that arise from a class of brain cells called astrocytes. The tumors can be classified by their level of malignancy on a grade scale from I – IV, with Grade I representing benign, non-cancerous tumors, and Grade IV (high-grade astrocytomas) representing highly invasive, very aggressive malignant tumors. These Grade IV astrocytomas are also referred to as glioblastoma multiforme, or GBM for short. As the disease progresses, the overall survival rate decreases dramatically from 8-10 years for patients diagnosed with Grade II tumors to 9-12 months for those diagnosed with GBM. Unfortunately, the majority (~70%) of low- and medium-grade astrocytomas go undetected only to emerge, without clinical warning, as GBM. In addition to the obvious tragedy this represents for patients and their family members, the dismally low detection rate of lower grade astrocytomas also means that researchers seeking to understand the progression of this disease are severely handicapped, lacking the crucial tissue samples from the stages that lie between health and advanced disease.

To address this challenge and gain insight into the mechanisms underlying the GBM development, researchers from across the country, including Burak Kutlu, Rhishikesh Bargaje and Lee Hood at ISB, used a group of mice that had been genetically modified to begin developing astrocytomas upon exposure to a specific chemical. Because the researchers would be able to control the disease initiating event, they could monitor both the transition from health to disease and the progression of the tumor from low-grade to high. This proved to be a powerful model for studying GBM development. Focusing in on three major networks that had previously been shown to be affected in GBM patients, the researchers were able to determine the exact order, from tumor initiation through progression to full-blown GBM, in which the networks malfunctioned. More specifically, they now know which of the three networks to examine for identifying biomarkers of early disease.

Lung Cancer

For a patient, determining whether a growth is benign can be as important, and as challenging, as detecting cancer. Often, diagnosis of a growth requires an invasive and costly surgical biopsy. Lung nodules are growths that, if malignant, may lead to lung cancer. A whopping 80 percent of patients who are diagnosed with having lung nodules endure invasive biopsies that turn out to be unnecessary. Researchers at ISB and Integrated Diagnostics (Indi is an ISB spinout) designed a blood test consisting of a panel of 13 biomarkers that will assist in identifying benign lung nodules. It has a 90% negative predictive value (NPV), which means that if the test calls the nodule benign (i.e. the result is negative), 90% of the time the test is correct. Indi currently is in the process of commercializing the diagnostics test, which ultimately will save patients from unnecessary — and expensive — invasive procedures.

Use of Clustering Methodology in Systems Biology

Clustering is the grouping of a particular set of objects based on their characteristics, aggregating them according to their similarities. Regarding to data mining, this methodology partitions the data implementing a specific join algorithm, most suitable for the desired information analysis.

This clustering analysis allows an object not to be part of a cluster, or strictly belong to it, calling this type of grouping hard partitioning. In the other hand, soft partitioning states that every object belongs to a cluster in a determined degree. More specific divisions can be possible to create like objects belonging to multiple clusters, to force an object to participate in only one cluster or even construct hierarchical trees on group relationships.

There are several different ways to implement this partitioning, based on distinct models. Distinct algorithms are applied to each model, differentiating it's properties and results. These models are distinguished by their organization and type of relationship between them. The most important ones are:

- Centralized: Each cluster is represented by a single vector mean, and a object value is compared to these mean values.
- Distributed: The cluster is built using statistical distributions.
- Connectivity: The connectivity on these models is based on a distance function between elements.
- Group: Algorithms have only group information.
- Graph: Cluster organization and relationship between members is defined by a graph linked structure.
- Density: Members of the cluster are grouped by regions where observations are dense and similar.

Broadly speaking, clustering can be divided into two subgroups:

- Hard Clustering: In hard clustering, each data point either belongs to a cluster completely or not. For example, in the above example each customer is put into one group out of the 10 groups.
- Soft Clustering: In soft clustering, instead of putting each data point into a separate cluster, a probability or likelihood of that data point to be in those clusters is assigned. For example, from the above scenario each costumer is assigned a probability to be in either of 10 clusters of the retail store.

Clustering

Clustering is the process of making a group of abstract objects into classes of similar objects. A cluster of data objects can be treated as one group. While doing cluster analysis, we first partition the set of data into groups based on data similarity and then assign the labels to the groups. The main advantage of clustering over classification is that, it is adaptable to changes and helps single out useful features that distinguish different groups.

Applications of Cluster Analysis

- Clustering analysis is broadly used in many applications such as market research, pattern recognition, data analysis, and image processing.
- Clustering can also help marketers discover distinct groups in their customer base. And they can characterize their customer groups based on the purchasing patterns.
- In the field of biology, it can be used to derive plant and animal taxonomies, categorize genes with similar functionalities and gain insight into structures inherent to populations.
- Clustering also helps in identification of areas of similar land use in an earth observation database. It also helps in the identification of groups of houses in a city according to house type, value, and geographic location.
- Clustering also helps in classifying documents on the web for information discovery.
- Clustering is also used in outlier detection applications such as detection of credit card fraud.
- As a data mining function, cluster analysis serves as a tool to gain insight into the distribution of data to observe characteristics of each cluster.

Requirements of Clustering in Data Mining

The following points throw light on why clustering is required in data mining:

- Scalability: We need highly scalable clustering algorithms to deal with large databases.
- Ability to deal with different kinds of attributes: Algorithms should be capable to be applied on any kind of data such as interval-based (numerical) data, categorical, and binary data.
- Discovery of clusters with attribute shape: The clustering algorithm should be capable of detecting clusters of arbitrary shape. They should not be bounded to only distance measures that tend to find spherical cluster of small sizes.
- High dimensionality: The clustering algorithm should not only be able to handle low-dimensional data but also the high dimensional space.
- Ability to deal with noisy data: Databases contain noisy, missing or erroneous data. Some algorithms are sensitive to such data and may lead to poor quality clusters.
- Interpretability: The clustering results should be interpretable, comprehensible, and usable.

Clustering Methods

Clustering methods can be classified into the following categories:

- Partitioning Method,
- Hierarchical Method,

- Density-based Method,
- Grid-Based Method,
- Model-Based Method,
- Constraint-based Method.

Partitioning Method

Suppose we are given a database of 'n' objects and the partitioning method constructs 'k' partition of data. Each partition will represent a cluster and $k \le n$. It means that it will classify the data into k groups, which satisfy the following requirements:

- Each group contains at least one object.
- Each object must belong to exactly one group.

Points to Remember:

- For a given number of partitions (say k), the partitioning method will create an initial partitioning.
- Then it uses the iterative relocation technique to improve the partitioning by moving objects from one group to other.

Hierarchical Methods

This method creates a hierarchical decomposition of the given set of data objects. We can classify hierarchical methods on the basis of how the hierarchical decomposition is formed. There are two approaches here:

- Agglomerative Approach: This approach is also known as the bottom-up approach. In this, we start with each object forming a separate group. It keeps on merging the objects or groups that are close to one another. It keep on doing so until all of the groups are merged into one or until the termination condition holds.
- Divisive Approach: This approach is also known as the top-down approach. In this, we start with all of the objects in the same cluster. In the continuous iteration, a cluster is split up into smaller clusters. It is down until each object in one cluster or the termination condition holds. This method is rigid, i.e., once a merging or splitting is done, it can never be undone.

Approaches to Improve Quality of Hierarchical Clustering

Here are the two approaches that are used to improve the quality of hierarchical clustering:

- Perform careful analysis of object linkages at each hierarchical partitioning.
- Integrate hierarchical agglomeration by first using a hierarchical agglomerative algorithm to group objects into micro-clusters, and then performing macro-clustering on the micro-clusters.

Density-Based Method

This method is based on the notion of density. The basic idea is to continue growing the given cluster as long as the density in the neighborhood exceeds some threshold, i.e., for each data point within a given cluster, the radius of a given cluster has to contain at least a minimum number of points.

Grid-Based Method

In this, the objects together form a grid. The object space is quantized into finite number of cells that form a grid structure. Advantages are:

- The major advantage of this method is fast processing time.
- It is dependent only on the number of cells in each dimension in the quantized space.

Model-Based Methods

In this method, a model is hypothesized for each cluster to find the best fit of data for a given model. This method locates the clusters by clustering the density function. It reflects spatial distribution of the data points. This method also provides a way to automatically determine the number of clusters based on standard statistics, taking outlier or noise into account. It therefore yields robust clustering methods.

Constraint-Based Method

In this method, the clustering is performed by the incorporation of user or application-oriented constraints. A constraint refers to the user expectation or the properties of desired clustering results. Constraints provide us with an interactive way of communication with the clustering process. Constraints can be specified by the user or the application requirement.

Clustering Algorithms in Data Mining

Based on the recently described cluster models, there are a lot of clustering that can be applied to a data set in order to partition the information. It is important to mention that every method has it's advantages and cons. The choice of algorithm will always depended on the characteristics of the data set and what we want to do with it.

- Centroid-based: In this type of grouping method, every cluster is referenced by a vector of values. Each object is part of the cluster whose value difference is minimal, comparing to other clusters. The number of clusters should be pre-defined, and this is the biggest problem of this kind of algorithms. This methodology is the most close to the classification subject and is vastly used for optimization problems.
- Distributed-based: Related to pre-defined statistical models, the distributed methodology combines objects whose values belong to the same distribution. Because of its random nature of value generation, this process needs a well-defined and complex model to interact in a better way with real data. However these processes can achieve an optimal solution and calculate correlations and dependencies.

- Connectivity-based: In this type of algorithm, every object is related to its neighbors, depending the degree of that relationship on the distance between them. Based on this assumption, clusters are created with nearby objects, and can be described as a maximum distance limit. With this relationship between members, these clusters have hierarchal representations. The distance function varies on the focus of the analysis.
- Density-based: These algorithms create clusters according to the high density of members of a data set, in a determined location. It aggregates some distance notion to a density standard level to group members in clusters. These kind of processes may have less performance on detecting the limit areas of the group.

Main Applications of Cluster Analysis

Since this is a very valuable data analysis technique, it has several different applications in the sciences world. Every large data set of information can be processed by this kind of analysis, producing great results with many distinct types of data.

One of the most important application is related to image processing detecting distinct kinds of pattern in image data. This can be very effective in biology research, distinguishing objects and identifying patterns. Another use is the classification of medical exams.

The personal data combined with shopping, location, interest, actions and a infinite number of indicators, can be analyzed with this methodology, providing very important information and trends. Examples of this are the market research, marketing strategies, web analytics, and a lot of others. Other types of applications based on clustering algorithms are climatology, robotics, recommender systems, mathematical and statistical analysis, providing a broad spectrum of utilization.

K Means Clustering

K means is an iterative clustering algorithm that aims to find local maxima in each iteration. This algorithm works in these 5 steps:

1. Specify the desired number of clusters K: Let us choose k=2 for these 5 data points in 2- D space.

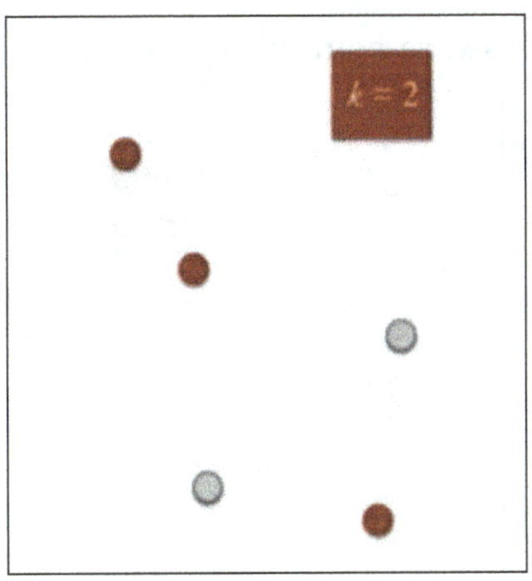

2. Randomly assign each data point to a cluster: Let's assign three points in cluster 1 shown using red color and two points in cluster 2 shown using grey color.

3. Compute cluster centroids: The centroid of data points in the red cluster is shown using red cross and those in grey cluster using grey cross.

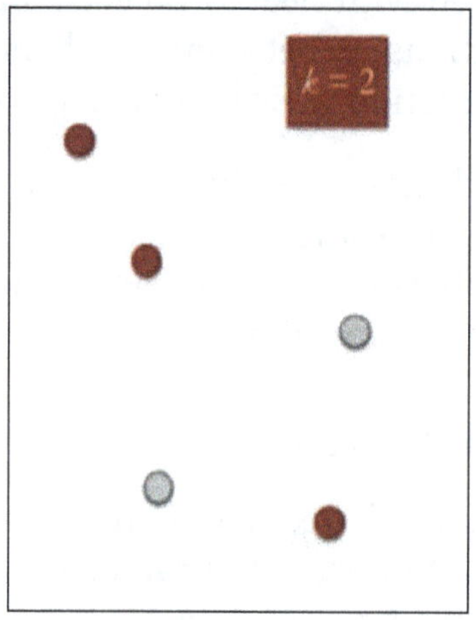

4. Re-assign each point to the closest cluster centroid: Note that only the data point at the bottom is assigned to the red cluster even though it is closer to the centroid of grey cluster. Thus, we assign that data point into grey cluster.

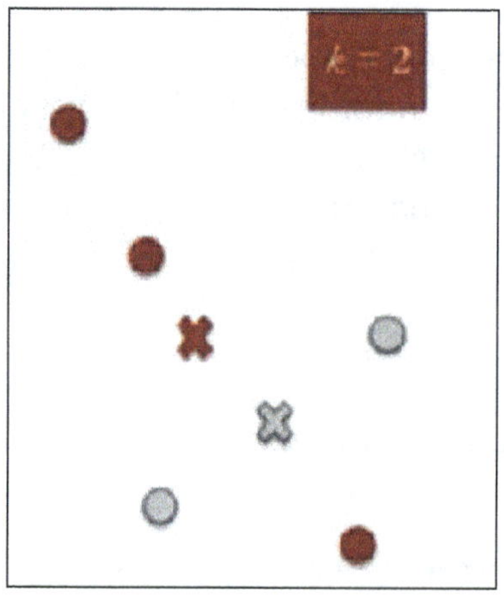

5. Re-compute cluster centroids: Now, re-computing the centroids for both the clusters.

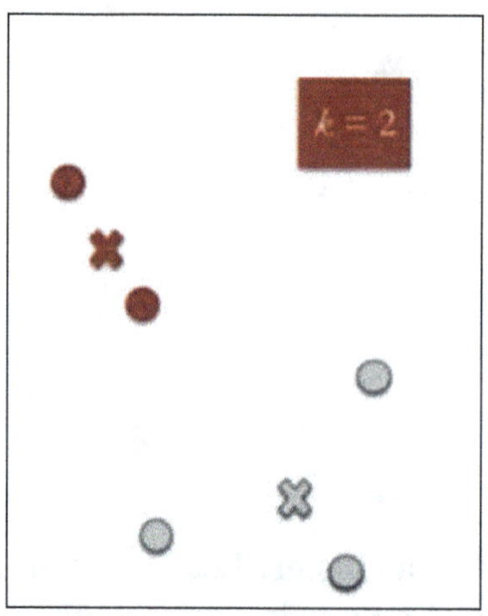

6. Repeat steps 4 and 5 until no improvements are possible: Similarly, we'll repeat the 4th and 5th steps until we'll reach global optima. When there will be no further switching of data points between two clusters for two successive repeats. It will mark the termination of the algorithm if not explicitly mentioned.

Hierarchical Clustering

Hierarchical clustering, as the name suggests is an algorithm that builds hierarchy of clusters. This algorithm starts with all the data points assigned to a cluster of their own. Then two nearest clusters are merged into the same cluster. In the end, this algorithm terminates when there is only a single cluster left. The results of hierarchical clustering can be shown using dendrogram. The dendrogram can be interpreted as:

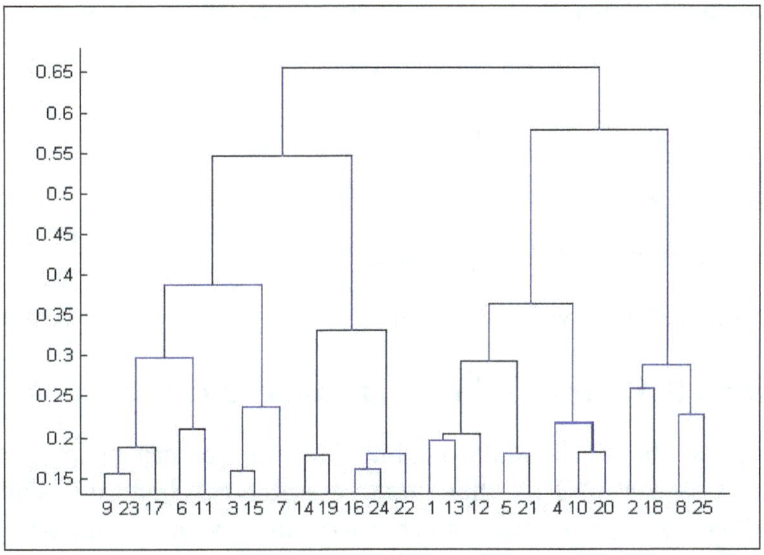

At the bottom, we start with 25 data points, each assigned to separate clusters. Two closest clusters are then merged till we have just one cluster at the top. The height in the dendrogram at which two clusters are merged represents the distance between two clusters in the data space.

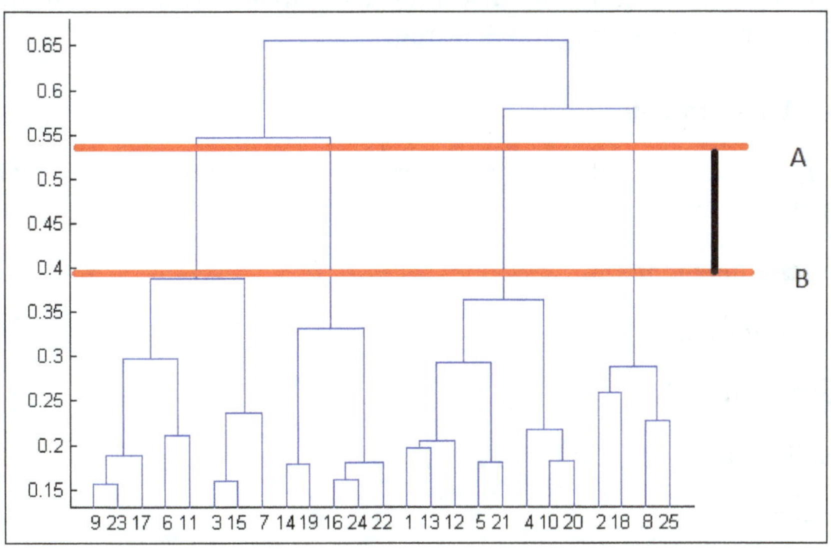

The decision of the number of clusters that can best depict different groups can be chosen by observing the dendrogram. The best choice of the number of clusters is the number of vertical lines in the dendrogram cut by a horizontal line that can transverse the maximum distance vertically without intersecting a cluster. In the above example, the best choice of no. of clusters will be 4 as the red horizontal line in the dendrogram below covers maximum vertical distance AB.

Two important things that we should know about hierarchical clustering are:

- This algorithm has been implemented above using bottom up approach. It is also possible to follow top-down approach starting with all data points assigned in the same cluster and recursively performing splits till each data point is assigned a separate cluster.
- The decision of merging two clusters is taken on the basis of closeness of these clusters. There are multiple metrics for deciding the closeness of two clusters:
    - Euclidean distance: $||a-b||_2 = \sqrt{(\Sigma(a_i-b_i))}$.
    - Squared Euclidean distance: $||a-b||_2^2 = \Sigma((a_i-b_i)^2)$.
    - Manhattan distance: $||a-b||_1 = \Sigma|a_i-b_i|$.
    - Maximum distance: $||a-b||_{INFINITY} = \max_i|a_i-b_i|$.
    - Mahalanobis distance: $\sqrt{((a-b)^T S^{-1} (-b))}$ {where, s : covariance matrix}.

Difference between K Means and Hierarchical Clustering

Hierarchical clustering can't handle big data well but K Means clustering can. This is because the time complexity of K Means is linear i.e. O(n) while that of hierarchical clustering is quadratic i.e. $O(n^2)$. In K Means clustering, since we start with random choice of clusters, the results produced by running the algorithm multiple times might differ. While results are reproducible in Hierarchical clustering.

K Means is found to work well when the shape of the clusters is hyper spherical (like circle in 2D, sphere in 3D). K Means clustering requires prior knowledge of K i.e. no. of clusters you want to divide your data into. But, you can stop at whatever number of clusters you find appropriate in hierarchical clustering by interpreting the dendrogram.

Applications of Clustering

Clustering has a large no. of applications spread across various domains. Some of the most popular applications of clustering are:

- Recommendation engines,
- Market segmentation,
- Social network analysis,
- Search result grouping,
- Medical imaging,

- Image segmentation,
- Anomaly detection.

Improving Supervised Learning Algorithms with Clustering

Clustering is an unsupervised machine learning approach, but can it be used to improve the accuracy of supervised machine learning algorithms as well by clustering the data points into similar groups and using these cluster labels as independent variables in the supervised machine learning algorithm? Let's find out.

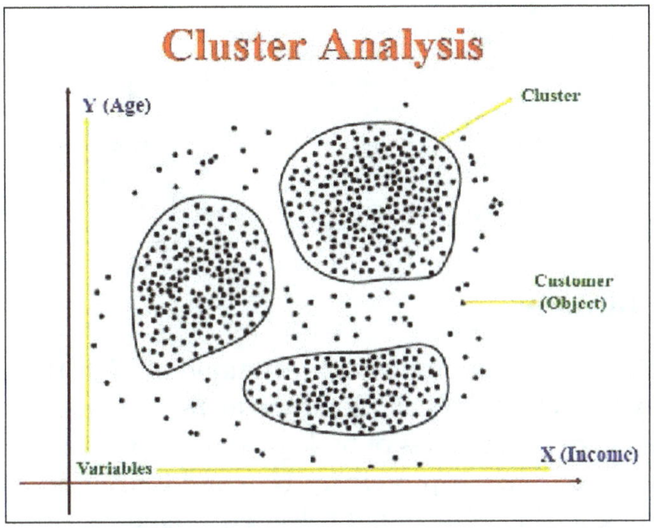

Mathematical Modeling in Systems Biology

Biological systems are inherently complex, and many of their emergent properties result from the interplay of numerous molecular components. Moreover, biochemical reactions often obey nonlinear reaction kinetics—that is, an increase in the amount of the starting material of the reaction does not necessarily lead to a proportional increase in the amount of the reaction product. Finally, other complexities, such as cell structure and compartmentalization or random (i.e., stochastic) effects, also often result in unexpected behavior of the entire system. Mathematical models that take these factors into consideration allow researchers to capture the features of complex biological systems and to understand how biological systems respond to external or internal signals and perturbations, such as different growth or development conditions or stress triggered by agents such as alcohol. Mathematical models have the big advantage of being amenable to computer simulations.

Models describing biological systems generally are too complex to be solved analytically ("manually") and therefore typically are solved numerically— that is, using computers to solve the mathematical equations that help predict the response of a biological system. With the availability of computer-based techniques for solving mathematical equations, the response of a biological system to different conditions can be relatively easily simulated in silico once a mathematical model is available. These computer simulations (so-called "dry experiments") in many cases require much lower investment and much less time compared with the typically more time-consuming and expensive biological experiments (sometimes referred to as "wet experiments").

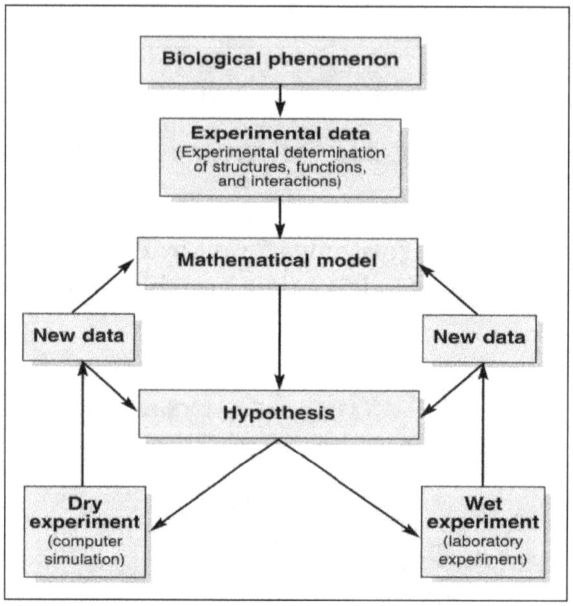

Network in System Biology

The development in molecular biology tools and technique of high-throughput techniques such as microarray technology availability of completely sequenced genomes and the and two-hybrid systems have allowed genome-wide studies of biological processes that arise from complex interaction between genetic entities, such as proteins, RNAs and DNA. These cellular entities interact with one another according to their functional roles. For instance, protein-protein interactions, related to cellular communication by signal transduction, activate or repress the transcription of genes, changing the molecular composition of the cell. These separate modules are 'building blocks' of the biological systems, working together to shape the phenotypic pattern of the cells and organisms. Mapping out all these interconnected modules helps to understand and model their topological and dynamical properties.

The most critical networks for controlling cellular systems are those related to transcription, protein-protein interaction and metabolism. Despite the great diversity of networks in cell biology, they share several global properties. For instance, most networks within the cell are characterized by (i) a scale-free degree distribution (also called power-law), (ii) a small average shortest path length between any two nodes (small-world), (iii) a disassortative nature, (iv) a modular organization and (v) a structural and dynamical robustness.

In the era of post-genomic biology, scientists have proposed the new and innovative idea that cells can be understood in terms of their network structure. To define the network structure of a cell, network biology researchers use data generated by experimental methods, including high-throughput proteomic, genomic, and metabolomic data, as well as computational capabilities to identify and map cell signaling networks. It has been shown that by identifying the most highly connected proteins, the overall network topology can be inferred. These interaction networks can be defined in terms of the protein complexes within cells and their relationships to each other.

Viewing cells in terms of their underlying network structure is a very powerful concept. All networks share common characteristics, and mathematical treatments have been developed to understand their structure and how they can be regulated. Thus, organizing biological information in the context of networks is fundamental to applying systems-level approaches to understanding biological function. By defining protein complexes as dynamic structures and quantifying the extent and quality of interactions of the partners, information relating to their function can be inferred.

Scope of Systems Biology

System Biology essentially involves:

- Methods and technologies for monitoring cell-wide gene expression.
- Mathematical and computational analysis of gene expression data.
- Methods and technologies for monitoring cell-wide protein expression.
- Mathematical and computational analysis of protein expression data.
- Methods and technologies for identification of protein-protein interaction.
- Mathematical and computational analysis of protein-protein interaction data.
- Methods and technologies for identification of DNA-protein interaction.
- Mathematical and computational analysis of DNA-protein interaction data.
- Genetic networks.
- Mathematical methods for the identification of genetic regulatory networks.
- Modeling and Simulation of gene expression networks.
- Translation networks.
- Modeling and Simulation of protein expression networks.
- Methods and technologies for monitoring metabolic reaction networks.
- Mathematical and computational analysis of metabolic reaction networks.

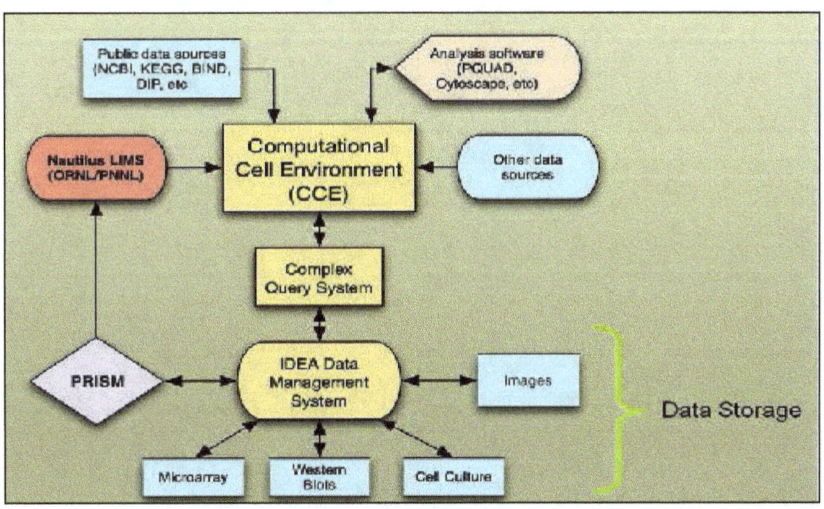

Data management in biological sciences.

Tools and methods for data collection, analysis, storage, and sharing are being integrated.

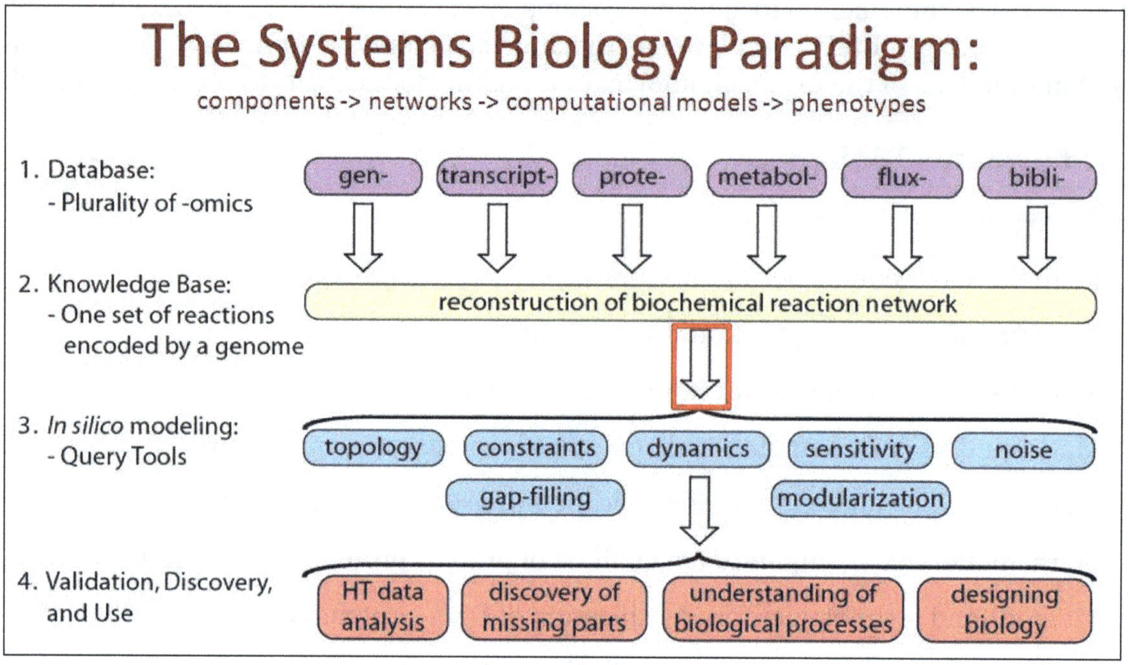

References

- Biophysics: biologydictionary.net, Retrieved 15, January 2020
- Quantum-biology-a-scientific-revolution-in-our-understanding-ofbiological-systems-2329-6577-1000185: longdom.org, Retrieved 03, August 2020

2

Cellular and Molecular Biophysics

Molecular biophysics is an interdisciplinary field of research that involves concepts of engineering, physics, chemistry, biology and mathematics to understand biomolecular systems. A few of the important topics studied under cellular and molecular physics are cellular function, cell division, cell cycle regulation machinery, etc. These topics are crucial for a complete understanding of the subject.

Cellular biophysics is the branch of biophysics that studies cells from the perspective of a physicist or physical chemist by applying physical methods to interrogate cell structure and function, and developing models of cells using physics and physical-chemical principles. Molecular biophysics is the study of the physical principles governing biomolecular systems. The goal of Molecular biophysics is to explain biological phenomena in terms of the physical and chemical properties of participating ions and molecules.

Biological life forms are composed of cells, which are the smallest unit of life, and then chemically can be broken down to molecules, which are different arrangements and combinations of atoms, the smallest particles which retain an element's identity. In a multicellular organism, these cells compose tissues, which compose organs, which compose organ systems, which ultimately form the entire organism. Different organisms all have the same basic biological structure, which starts with the cell. Single-celled organisms, while only being composed of a single cell, are still very complex in their intracellular workings. Multi-cellular organisms, such as ourselves, however, are much more complex and are composed of many different types of cells which all work together.

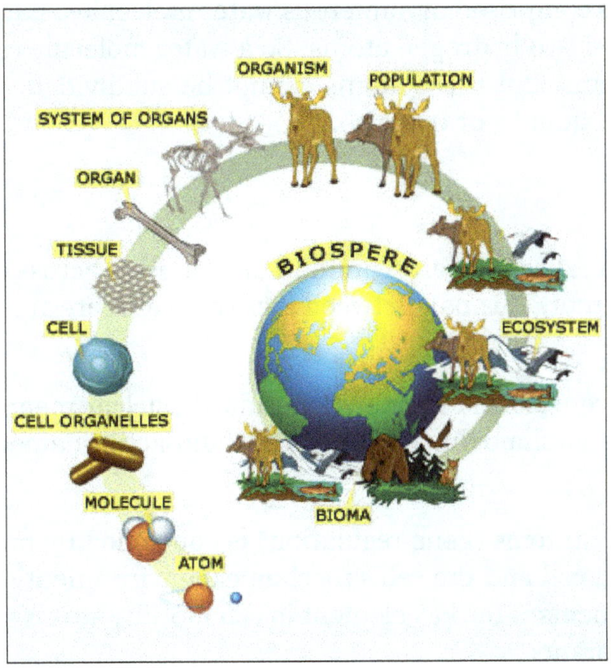

- Molecules: Group of atoms.
- Cells: Smallest unit of living things.
- Tissues: Groups of cells working together.
- Organs: Groups of tissues working together.
- Organ Systems: Groups of organs working together.
- Organism: Organ systems working together.
- Population: Group of organism in the same area.
- Community: Populations that live together in a defined area.
- Ecosystem: Community and its surroundings.
- Biosphere: The part of the Earth that contains ecosystems.

Atom

An atom is smallest particle in an element that has the properties of the element. It is not possible to breakdown the atom further retaining the properties of the element. Atoms are not visible to the naked eye and are the basic building blocks. For example the atoms of element gold cannot be broken down further and each atom has the properties of gold.

Molecules

Molecules are formed by the combination of two or more atoms. Unlike atoms, molecules can be subdivided to individual atoms. The atoms are bonded together in a molecule. Molecules also are not visible to the naked eye, while can be seen through highly magnifying microscopes and other scientific devices. Water is comprised of numerous water molecules. Each water molecule is made up of one oxygen atom and two hydrogen atoms. So a water molecule can be further divided into oxygen and hydrogen atoms. But these atoms cannot be subdivided. In a molecule, atoms are bonded together by single, double, or triple bonds.

Cells

Cells are a basic unit of life, sharing many common mechanisms across different classes of organisms (plants, animals, microorganisms). Between these classes there are also key differences in cell architecture.

In a cell, the external bilayer (plasma membrane) and the nuclear membrane are built and maintained by lipids. Stiff protein filaments (microtubules and actin filaments) confer rigidity to the cell's cytoskeleton.

The rate of production of proteins (gene regulation) is controlled by many factors, including the physical forces acting on a cell and the cell's mechanical environment. In animal cells a network (polymer gel) of actin filaments is the key element in cell motility as it is stiff enough to exert forces that deform the cell membrane.

All living organisms are made of cells, which are the units of life. A cell comprises a plasma membrane consisting of a phospholipid bilayer, genetic material, and various organelles that control the cellular function and respond to various stimuli.

Cellular Function

A cell's function is achieved through the culmination of hundreds of smaller processes, many of which are dependent on each other, and share protein or molecular components. Despite the phenotypic and functional variations that exist between cell types, it remains true that a high level of similarity exists when exploring subcellular processes, the components involved, and importantly, the organization of these components.

Importance of Compartmentalization

Cells are not an amorphous mixture of proteins, lipids and other molecules. Instead, all cells are comprised of well-defined compartments, each specializing in a particular function. In many cases subcellular processes may be described based on whether they occur at the plasma membrane, within the cytosol or within membrane bound organelles such as the nucleus, Golgi apparatus or even vesicular components of the membrane trafficking system like lysosomes and endosomes. Compartmentalization can also have important physiological implications.

Spatiotemporal Control of Various Macromolecules

In eukaryotes, synthesis of DNA, RNA, proteins and lipids is performed in a spatiotemporal manner. Each molecule is produced within specialized organelles or compartments with strict regulatory mechanisms existing to control the timing and rate of synthesis.

Example: Like DNA and RNA are synthesized in the nucleus, while Protein is synthesized in the cytoplasm. In prokaryotes, where there are no separate compartments, both transcription and translation occur simultaneously.

Protein Localization

In order for subcellular processes to be carried out within defined compartments or cellular regions, mechanisms must exist to ensure the required protein components are present at the sites and at an adequate concentration. The accumulation of a protein at a given site is known as protein localization.

Recruitment of Protein to specific organelles is essentially a form of protein recognition, made possible by the presence of specific amino-acid sequences within the protein structure. For example, polarized epithelial cells, which possess distinct apical and basolateral membranes, can, for instance produce a secretory surface for various glands. Similarly, neuronal cells develop effective networks due to the production of dendrites and axonal processes from opposite ends of the cell body. Moreover, in the case of embryonic stem cells, cell polarization can result in distinct fates of the daughter cells.

Directed Delivery of Components

Protein localization can result from the recognition of passively diffusing soluble proteins or protein complexes; however, this may not guarantee a sufficient concentration of components to maintain a given process. This can impede its completion, particularly when carried out in regions with a limited cytoplasmic volume, such as the tip of a filopodia, or when components are rapidly turned over. A more efficient way of maintaining the concentration of protein components is by their directed delivery via the cytoskeletal network.

Communication Pathways

With different processes being carried out in separate subcellular compartments, organized across different regions of the cell, intracellular communication is paramount, which is achieved through various signaling pathways. This ultimately ensures individual compartments function efficiently, and enables one subcellular process to drive another and allows a cell to facilitate its primary functions in an efficient and coherent manner.

Difference in Cellular Organization

Prokaryotes:

- Have no membrane-bound organelles. Have genetic material but no nucleus.
- Include bacteria, fungi and algae.

Eukaryotes:

- Have membrane bound organelles.
- Includes animal and plants.

Cell Organelles and their Functions

Organelles make up the subunits of a cell. There are numerous each with their own function.

Plasma Membrane

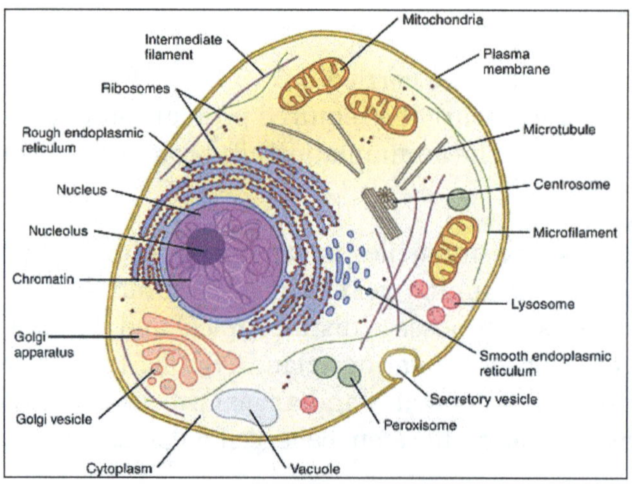

Animal cell showing Organelles.

The plasma membrane is the organelle that encapsulates the contents of the cell. Apart from encapsulating cell contents, the plasma membrane also plays a vital role in regulating the movement of substances in and out of the cell.

As such, it is actively involved in such both passive and active transportation to and from the cell. These processes also help maintain balance even when conditions outside the cell change. The plasma membrane is made up of two layers of phospholipids (phospholipids bilayer).

Nucleus/DNA

Some of the main components of the nucleus include the chromatic, nucleoplasm/nuclear sap and the nucleolus. The nucleus houses DNA (the hereditary material) as well as various proteins and the nucleolus.

In eukaryotic cells, the nucleus is enclosed in a nuclear membrane. It is the organelle that controls the hereditary traits of an organism by directing such processes as protein synthesis and cell division among others. For prokaryotes, the DNA lacks a nuclear membrane. The genetic material is therefore bound in the nucleotide region. The nucleolus plays an important role in ribosome production.

Ribosome

Ribosomes are tiny organelles that contain RNA and specific proteins within the cytoplasm. Within the cell, ribosomes are directly involved in the manufacture of proteins by using their RNA and amino acids. This process involves decoding the information contained in the mRNA and using amino acids to produce the required proteins.

Mitochondria

Mitochondria are some of the largest organelles within a cell. Compared to some of the other organelles, mitochondria contain DNA which makes them semiautonomous. Mitochondria also contain a double-membrane with the inner membrane folding to form cristae.

Also known as the powerhouse, mitochondria play an important role in respiration where they generate ATP (adenosine triphosphate) from substrates in the presence of oxygen. Using their DNA, mitochondria are able to encode for some of the components they require to perform their functions. ATP stores energy in the form of chemical bonds and is released whenever it is needed for various cell functions.

Vacuoles

A vacuole may be described as a space inside the cell that does not contain cytoplasm. It is surrounded by a membrane and filled with a fluid. Vacuoles store various molecules including enzymes, waste products of the cell, water, and even food material depending on the type of cell.

In cases where vacuoles contain waste products of the cell, they are also involved in the exportation of waste from the cell thus protecting the cell from toxicity. Some vacuoles also play a role in maintaining the internal hydrostatic pressure of the cell as well as regulating pH.

Cytoskeleton

The cytoskeleton is made up of microtubules and microfilaments. By spreading throughout the cell (in the cytoplasm), the cytoskeleton helps maintain the shape of the cell while also ensuring its elasticity. The cytoskeleton is also involved in anchoring the nucleus and supporting cell contents.

Plastids

Plastids are a type of organelle found in plant cells and algae. Like mitochondria, plastids are membrane-bound organelles that contain nucleoids. As such, they are also semi-autonomous organelles. There are different types of plastids that include chloroplasts, chromoplasts, gerontoplasts, and leucoplasts. The following are some functions of the different types of plastids:

- Chloroplast: Chloroplasts contain the pigment chlorophyll that captures energy from sunlight for photosynthesis. Therefore, the chloroplast is the site of photosynthesis (the process through which plants produce food).
- Chromoplasts: Chromoplasts are present in certain photosynthetic eukaryotes. They are primarily involved in the production and storage of pigments (carotenoid pigments). These pigments are involved in absorbing light energy as well as protecting chlorophyll in some plants.
- Gerontoplasts: This is a type of plastid that develops from the chloroplast during senescence in foliage plants. Gerontoplasts play an important role in salvaging nutrients and other important material as a cell dies.
- Leucoplast: Compared to the other plastids, leucoplasts are unpigmented organelles. Leucoplast is specialized for such functions as the storage of starch (amyloplasts), lipids (elaioplasts) and proteins (proteinoplasts).

Endoplasmic Reticulum

Found in eukaryotic cells, Endoplasmic reticulum (ER) is the organelle that forms an interconnected network of flattened sacs (cisternae). Like some of the other organelles found in eukaryotes, ER is enclosed in a membrane. The ER is divided into two regions that vary in structure and function. These include:

1. Smooth endoplasmic reticulum: The smooth ER is named so because it lacks a ribosome on its surface. As a result, it is smoother in appearance as compared to the rough ER. It is involved in the synthesis of lipids (e.g. phospholipids) and carbohydrates that are used to build the cell membrane. Some of the other functions of the smooth ER include:

- Transportation of vesicles.
- Enzyme production in the liver.
- Contraction of muscle cells in the muscles.
- Synthesis of hormones in the brain cells.

2. Rough endoplasmic reticulum: Unlike the smooth ER, rough ER has ribosome attached to its surface. It's involved in the manufacture of various proteins in the cell. On the other hand, the rough ER is involved in the production of antibodies, insulin as well as transportation of proteins into the smooth ER.

Centriole

Centrioles are cylindrical organelles found in most eukaryotic cells. They contain tube-shaped molecules known as microtubules that help separate chromosomes and move them during cell division.

Lysosome

A lysosome is commonly referred to as sacs of enzymes. They are membranous organelles that contain acidic enzymes (hydrolase enzymes) that serve to digest various macromolecules (e.g. lipids and nucleic acids) in the cell.

Conditions inside lysosomes have been shown to be acidic. These conditions are maintained by the lysosome membrane thus providing favorable conditions for the enzymes to perform their functions. Acidocalcisomes of eukaryotes are considered lysosome-related organelles.

Golgi Apparatus

Golgi apparatus are found in eukaryotic and are highly folded into cisternae (flattened sacs). They are enclosed in a membrane that varies in thickness from different regions. In the cell, Golgi apparatus are actively involved in the manufacturing, storage as well as transportation of products from the ER. Other features of a cell include:

- Cell Wall: It's one of the most important components of plant cells. The cell wall surrounds the cell membrane and serves to strengthen and protect the cell. For instance, in the cells of plant roots, the cell wall protects the cell as they grow deeper in the soil. The cell wall also serves as a filter that controls the movement of molecules in and out of the cell.
- Cytoplasm: It is an important component of the cell. Cell cytoplasm is composed of protoplasm in which all the other cell organelles are suspended. Many of the cell processes (protein synthesis, respiration etc.) take place in the cytoplasm. The cytoplasm also plays an important role in the movement of various materials around the cell.

Cell Division

Cell division is the process cells go through to divide. There are several types of cell division, depending upon what type of organism is dividing. Organisms have evolved over time to have different and more complex forms of cell division. Most prokaryotes, or bacteria, use binary fission to divide the cell. Eukaryotes of all sizes use mitosis to divide. Sexually-reproducing eukaryotes use a special form of cell division called meiosis to reduce the genetic content in the cell. This is necessary in sexual reproduction because each parent must give only half of the required genetic material, otherwise the offspring would have too much DNA, which can be a problem. These different types of cell division are discussed below.

Types of Cell Division

Prokaryotic Cell Division

Prokaryotes replicate through a type of cell division known as binary fission. Prokaryotes are simple organism, with only one membrane and no division internally. Thus, when a prokaryote divides, it simply replicates the DNA and splits in half. The process is a little more complicated than this, as DNA must first be unwound by special proteins. Although the DNA in prokaryotes usually exists in a ring, it can get quite tangled when it is being used by the cell. To copy the DNA efficiently, it must be stretched out. This also allows the two new rings of DNA created to be separated after they are produced. The two strands of DNA separate into two different sides of the prokaryote cell. The cell then gets longer, and divides in the middle. The process can be seen in the image below:

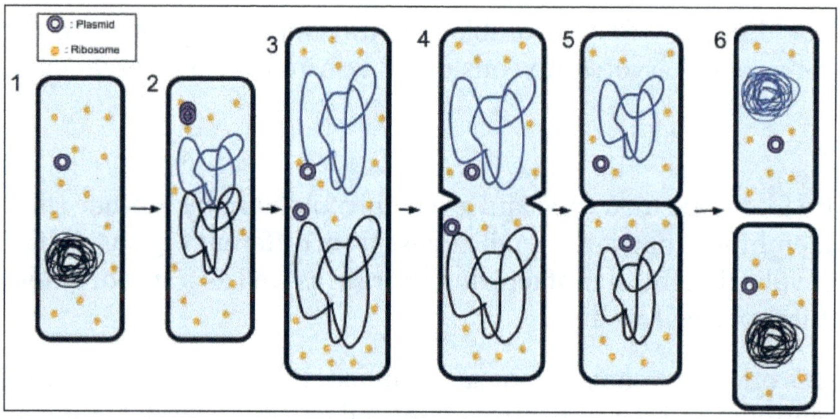

The DNA is the tangled line. The other components are labeled. Plasmids are small rings of DNA that also get copied during binary fission and can be picked up in the environment, from dead cells that break apart. These plasmids can then be further replicated. If a plasmid is beneficial, it will increase in a population. This is in part how antibiotic resistance in bacteria happens. The ribosomes are small protein structures that help produce proteins. They are also replicated so each cell can have enough to function.

Eukaryotic Cell Division

Mitosis

Eukaryotic organisms have membrane bound organelles and DNA that exists on chromosomes, which makes cell division harder. Eukaryotes must replicate their DNA, organelles, and cell mechanisms before dividing. Many of the organelles divide using a process that is essentially binary fission, leading scientist to believe that eukaryotes were formed by prokaryotes living inside of other prokaryotes.

After the DNA and organelles are replicated during interphase of the cell cycle, the eukaryote can begin the process of mitosis. The process begins during prophase, when the chromosomes condense. If mitosis proceeded without the chromosomes condensing, the DNA would become tangled and break. Eukaryotic DNA is associated with many proteins which can fold it into complex structures. As mitosis proceeds to metaphase the chromosomes are lined up in the middle of the cell. Each half of a chromosome, known as sister chromatids because they are replicated copies of each other, gets separated into each half of the cell as mitosis proceeds. At the end of mitosis, another process called cytokinesis divides the cell into two new daughter cells.

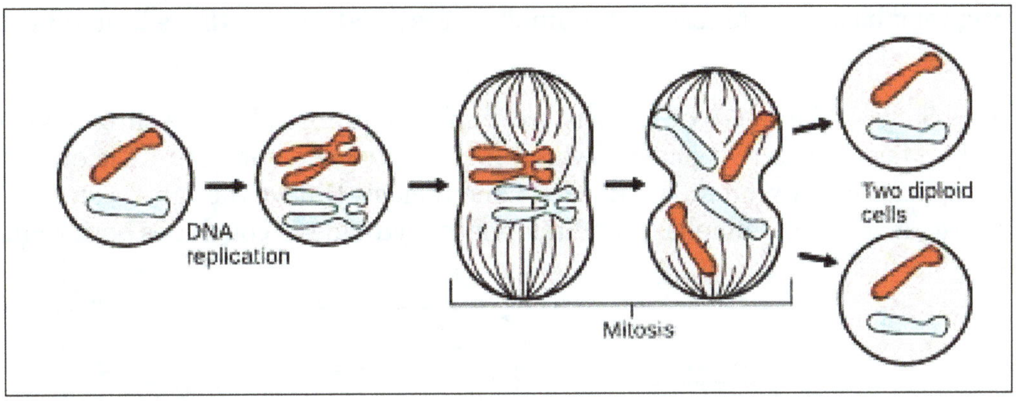

All eukaryotic organisms use mitosis to divide their cells. However, only single-celled organisms use mitosis as a form of reproduction. Most multicellular organisms are sexually reproducing and combine their DNA with that of another organism to reproduce. In these cases, organisms need a different method of cell division. Mitosis yields identical cells, but meiosis produces cells with half the genetic information of a regular cell, allowing two cells from different organisms of the same species to combine.

Meiosis

In sexually reproducing animals, it is usually necessary to reduce the genetic information before fertilization. Some plants can exist with too many copies of the genetic code, but in most organisms it is highly detrimental to have too many copies. Humans with even one extra copy of one chromosome can experience detrimental changes to their body. To counteract this, sexually reproducing organisms undergo a type of cell division known as meiosis. As before mitosis, the DNA and organelles are replicated. The process of meiosis contains two different cell divisions, which happen back-to-back. The first meiosis, meiosis I, separates homologous chromosomes. The homologous chromosomes present in a cell represent the two alleles of each gene an organism has. These alleles are recombined and separated, so the resulting daughter cells have only one allele for each gene, and no homologous pairs of chromosomes. The second division, meiosis II, separated the two copies of DNA, much like in mitosis. The end result of meiosis in one cell is 4 cells, each with only one copy of the genome, which is half the normal number.

Organisms typically package these cells into gametes, which can travel into the environment to find other gametes. When two gametes of the right type meet, one will fertilize the other and produce a zygote. The zygote is a single cell that will undergo mitosis to produce the millions of cells necessary for a large organism. Thus, most eukaryotes use both mitosis and meiosis, but at different stages of their lifecycle.

Cell Division Stages

Depending upon which type of cell division an organism uses, the stages can be slightly different:

- Mitosis Stages: Mitosis starts with prophase in which the chromosome is condensed. The cell proceeds to metaphase where the chromosomes are aligned on the metaphase plate. Then the chromosomes are separated in anaphase and the cell's cytoplasm is pinched apart during telophase. Cytokinesis is the final process that breaks the cell membrane and divides the cell into two.
- Meiosis Stages: The stages of meiosis are similar to mitosis, but the chromosomes act differently. Meiosis has two phases, which include two separate cell divisions without the DNA replicating between them. Meiosis I and meiosis II have the same 4 stages as mitosis: prophase, metaphase, anaphase, and telophase. Cytokinesis concludes both rounds of meiosis.

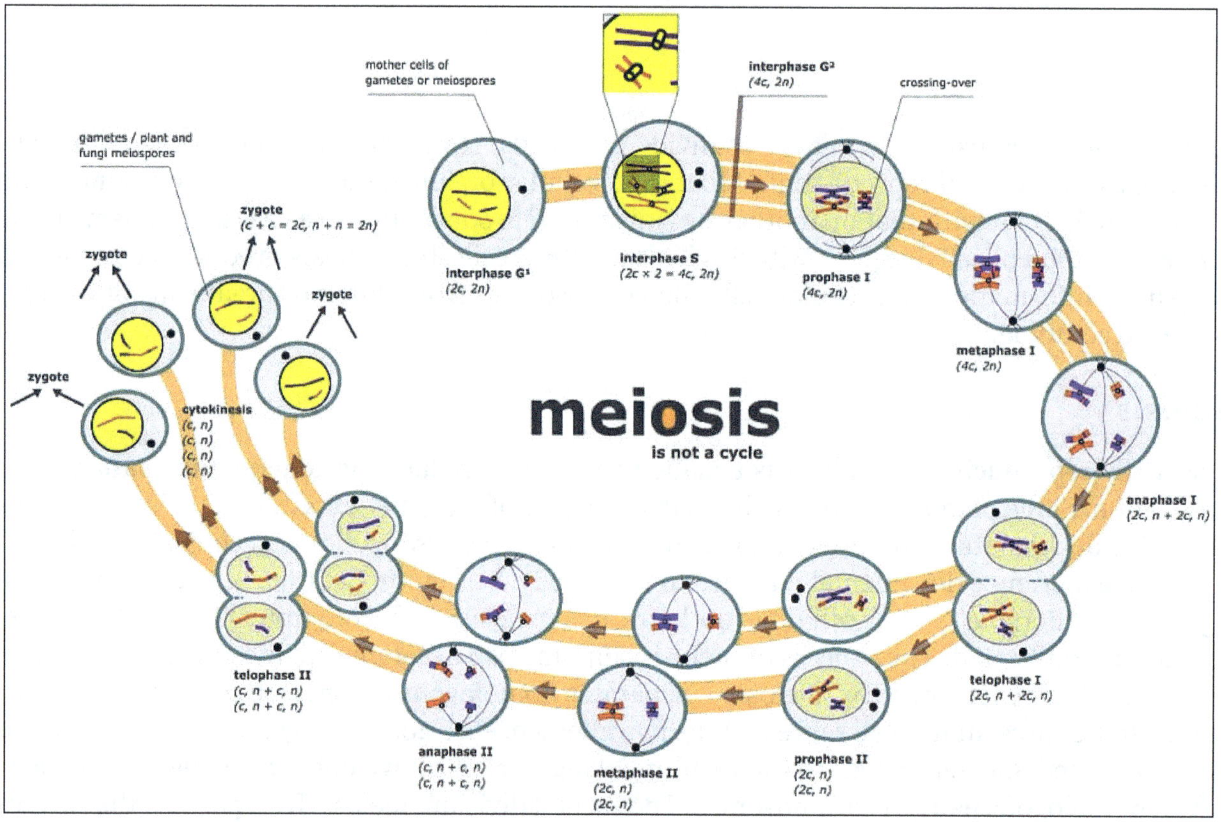

In prophase I, the chromosomes are condensed. In metaphase I, the chromosomes line up across from their homologous pairs. When they are separated in anaphase I and telophase I, there is only one form of each gene in each cell, known as a reduction division. Meiosis II proceeds in the same manner as mitosis, which sister chromatids dividing on the metaphase plate. By telophase II, there are 4 cells, each with half of the alleles as the parent cell and only a single copy of the genome. The cells can now become gametes and fuse together to create new organisms.

Cell Cycle Regulation Machinery

Cell cycle regulation machinery functions in all eukaryotic organisms to control the cell proliferation, while cancer is a disease of uncontrolled cellular proliferation. Proper functioning of the cell cycle regulation machinery plays the central role in preventing cells from becoming cancerous. One or the other component of cell cycle regulation is defective or mutated in most of the known cancers.

During the embryonic growth and later development, cell division takes place in virtually every tissue. In adult organisms most of the cells are differentiated and become quiescent. Whether a cell will undergo division or not, is of crucial importance and depends on the complex system of growth signals and the cell cycle regulatory mechanisms. The eukaryotic cell cycle occurs in four well-defined stages:

- S (synthetic) phase: The DNA replication takes place and the amount of DNA doubles in the cell.
- G_2 (Gap) phase: Proteins essential for cell division are synthesized and the cell size approximately doubles.
- M (mitosis) phase: Maternal nuclear envelope breaks down, paired chromosomes are pulled to opposite poles of the cell, and nuclear division takes place followed by cell division (cytokinesis), producing two daughter cells.
- G_1 (Gap) phase: This phase signifies a waiting period before the daughter cells can undergo cell division again which happens in embryonic and rapidly proliferating tissues. After passing through M phase and into G_1, a cell can continue to divide and produce daughter cells or it ceases to divide and enter a quiescent (G_0) phase, which may last hours, days, or even the lifetime of the cell. When a quiescent cell gets signals to divide again, it reenters the cell cycle through G_1 phase.

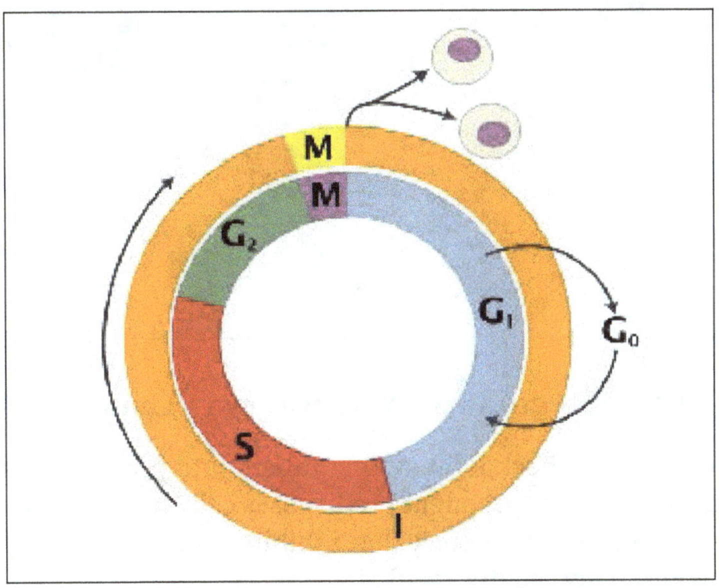

Cyclin and Cyclin Dependent Kinases (CDKs)

A cascade of protein phosphorylations relay the progression of cells through one stage of cell cycle to the next. This system involves highly regulated cyclin dependent kinases (CDKs) which are activated only after their association with another subunit called cyclin. Different CDKs associate with their corresponding cyclin subunits, which are transiently expressed at the appropriate period of the cell cycle to form an active complex with unique substrate specificity. Activated CDKs in these complexes drive the cell cycle by phosphorylating key proteins like Rb that regulate the cell cycle transitions.

CDK Inhibitors (CDKIs)

The activity of CDK-cyclin complexes is controlled by CDK inhibitors (CDKIs), which constitute cell cycle checkpoints. The surveillance mechanism senses the damage in DNA and chromosomes. The checkpoints ensure that cells with defective DNA or chromosomes do not replicate. The G1-S checkpoint monitors the integrity of DNA before its replication, while G2-M checkpoint checks DNA after replication and ensures that cell is suitable to enter mitosis. When DNA damage is detected, the checkpoint activation halts the progression of cell cycle and provides time for DNA repair machinery to function. If an irreparable damage to DNA has occurred, p53 mediated responses lead to elimination of cell by apoptosis or it enters a non-replicative state called senescence.

- Non-specific CDKIs- p21, p27 and p57.
- Specific cyclin D-CDK4/6 inhibitors- p15, p16, p18 and p19 collectively called as INK4 (A to D) proteins.

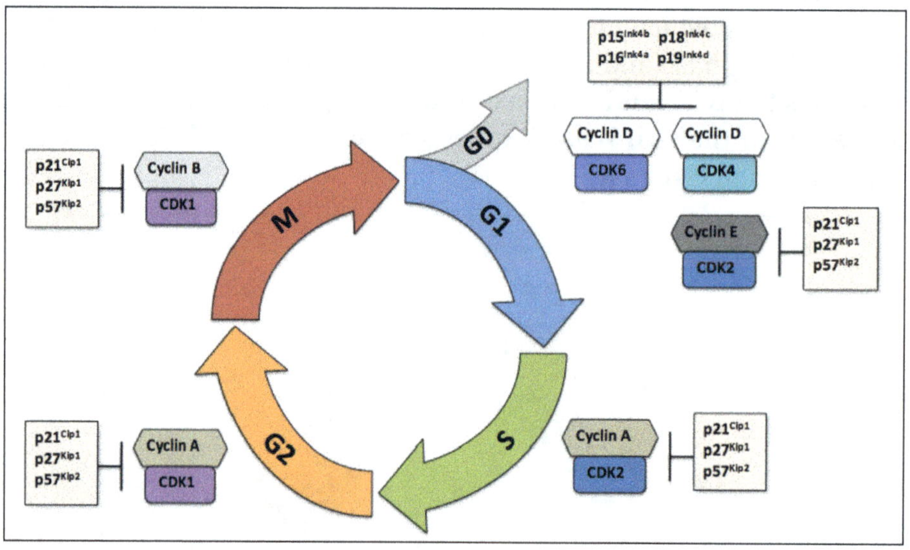

Alterations in Cell Cycle Regulation Machinery in Cancer Cells

The whole cell cycle control and checkpoint system function to prevent cells from becoming cancerous and alteration in any of the component predisposes cells for development of tumor. Almost all the cancers appear to have genetic defect that inhibits proper functioning of the G1-S checkpoint, causing cells to continually reenter the S phase.

Alterations in Activity of Cyclins, CDKs or CDKIs

- Cyclin D genes are overexpressed in many cancers including breast, esophagus, liver and certain type of lymphomas and plasma cell tumors.
- CDK4 gene overexpression is reported in melanomas, sarcomas and glioblastomas.
- Inhibition of CDKI (p16) activity is seen in many malignancies including some subsets of melanomas, pancreatic cancers, glioblastomas, esophageal cancer, lung cancers, soft tissue sarcomas and bladder cancers. MYC, a transcription factor, overexpressed in many cancers, is implicated in repressing the activity of CDKIs.

RB Gene: Governor of the Cell Cycle

The product of RB gene is a DNA binding protein which regulates G1-S transition in cell cycle. Inside the cells Rb is present in an active hypophosphorylated form and an inactive hyperphosphorylated form. In its active hypophosphorylated form, Rb binds and inactivates E2F family of transcription factors; preventing transcription of cyclin E. Active Rb blocks E2F mediated transcription in following two ways:

- Rb sequesters E2F, preventing it from interacting with other transcriptional activators.
- Rb recruits chromatin remodeling proteins like histone deacetylases and histone methyltransferases, which inhibit transcription of E2F responsive genes like cyclin E.

When cells receive mitogenic stimuli, cyclin D is expressed and activated cyclin D-CDK-4/6 complex is formed which inactivates Rb by phosphorylation and E2F is released leading to cyclin E expression. Cyclin E expression stimulates DNA replication and progression through the cell cycle.

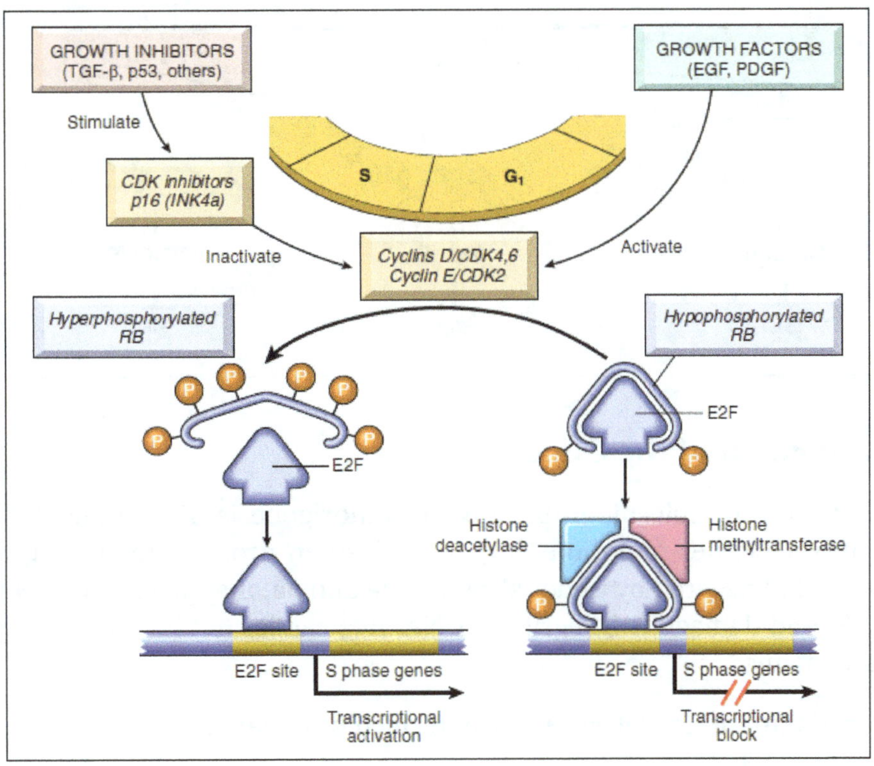

The loss of RB gene was first discovered in retinoblastomas but now it is evident that homozygous loss of RB gene is a common feature of several tumors including breast, bladder and small cell lung cancer. Patients of familial retinoblastoma are also at increased risk of developing osteosarcoma and certain soft tissue sarcomas. A cell heterozygous at the RB locus is not neoplastic. Tumors develop when the cell loses its normal RB gene copy and thus becomes homozygous for the mutant allele.

In cervical cancers Rb function is eliminated by the binding of an inhibitory protein, designated E7, which is encoded by human papillomavirus (HPV).

Mutation at INK4b-ARF-INK4a Locus

The locus encoding p16 is most highly vulnerable genetic locus in human genome. This locus encodes for three tumor suppressor genes. P16 is encoded by gene INK4a, immediately upstream of which is INK4b locus coding for p15, another cyclin D-CDK4/6 inhibitor. P14ARF is encoded by an exon upstream of the first INK4a exon and shares exon 2 and exon 3 with INK4a. p14ARF controls the stability of p53 and acts as a positive regulator of p53. The p14ARF gene is induced by high levels of mitogenic signaling. p14ARF binds and sequesters Mdm2 in nucleolus, where it cannot access p53 thereby playing an important role in stabilization of p53 by controlling levels of Mdm2. Thus mutations in this locus would simultaneously affect two major tumor suppressor pathways in the cell, the Rb and p53 pathways, resulting in carcinogenesis.

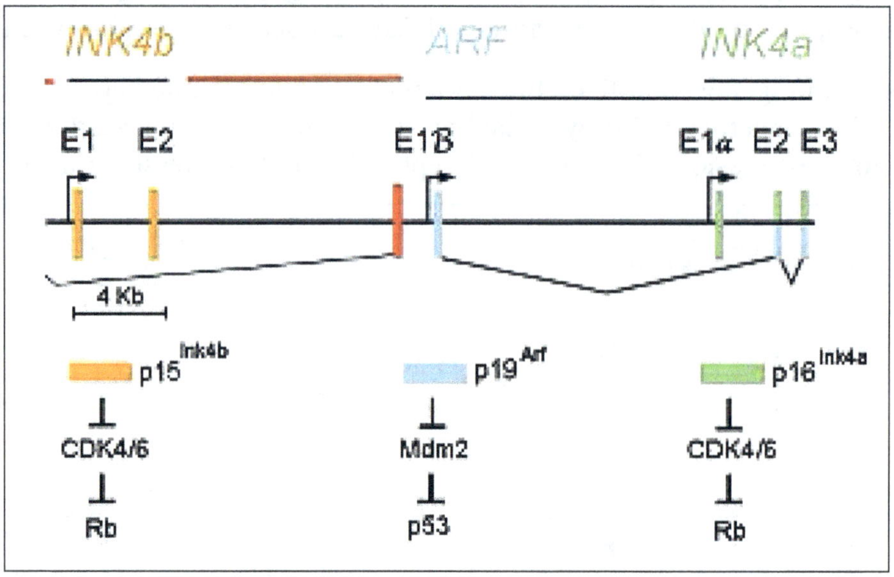

TP53 Gene: Guardian of the Genome

p53 is the most important molecule implicated in tumorigenesis. It is thought that most of the human cancers involve either mutations in p53 itself or in proteins regulating its activity. p53 response pathways are triggered by under stresses like anoxia, inappropriate oncoprotein activity (e.g., MYC or RAS) and damage to integrity of DNA. p53 carries out its tumor suppressor activity by following mechanisms:

- Activation of temporary cell cycle arrest, termed quiescence.

- Induction of permanent cell cycle arrest, termed senescence.
- Induction of programmed cell death, termed apoptosis.

In a healthy cell, the activity of p53 is normally kept low by a protein called Mdm2. The activated Mdm2 protein forms a complex with p53, inhibiting the transcription factor and ubiquitinating it for proteasomal degradation. The protein kinase activity of ATM and/or ATR is activated following DNA damage is sensed. ATM inactivates Mdm2 by phosphorylating it thereby increasing the half-life of p53 and activating its ability to drive transcription of various target genes.

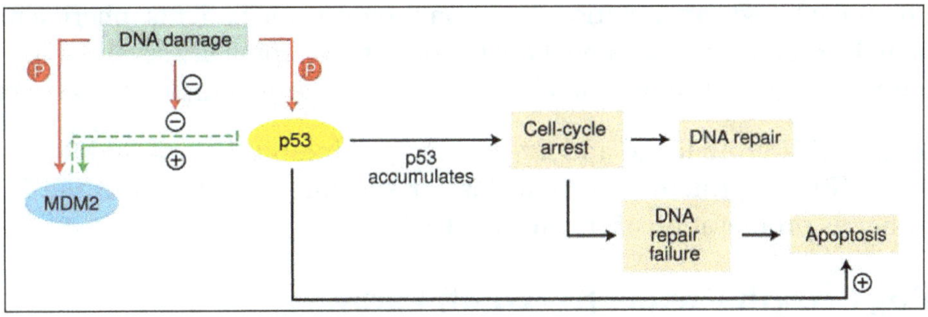

p53 mediated G1 cell cycle arrest is result of p53 dependent transcription of CDKI gene CDKN1A (p21). p21 inhibits cyclin-CDK complex and prevents phosphorylation of Rb, thereby arresting cells in G1 phase. The p53 protein also induces expression of DNA damage repair genes.

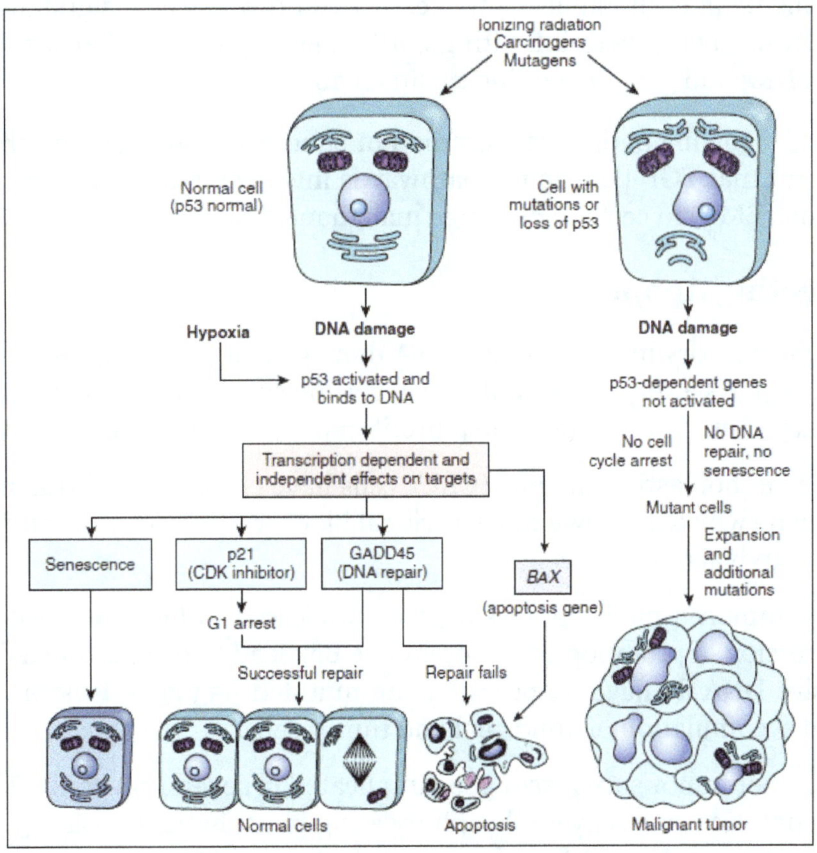

p53 acts as a "guardian of genome" and guards the cell from genetic changes. When p53 mediated

G1 checkpoint does not function properly, defective DNA can replicate, perpetuating mutations and DNA rearrangements that are passed on to daughter cells. This can result in transformation into cancerous and metastatic cells. p53 loss of function mutations can also result in inhibition of apoptosis, contributing to transformed cells. More than 70% of human cancers have a defect in TP53 gene including breast, colon and lung cancers.

The active form of p53 is a tetramer of four identical subunits. A missense mutation in one of the two p53 alleles leads to a loss of function because virtually all the oligomers will contain at least one defective subunit, and such oligomers have reduced ability to activate transcription. Such a mutation is an example of dominant negative mutation. Less commonly, some patients inherit a mutant copy of TP53; resulting in Li-Fraumeni syndrome. Patients of this syndrome are several folds more susceptible for developing malignant tumors and also at a younger age as compared to general population.

In cervical cancers, HPV proteins E6 and E7- bind to and inhibit the p53 and Rb tumor suppressors respectively. HPV E5 protein is responsible for sustained activation of the PDGF receptor leading to continuous proliferation of transformed cells.

Transforming Growth Factor–β (TGF-β) Pathway

TGF-β is a member of a family of dimeric growth factors which includes bone morphogenetic proteins and activins. In most of the normal cells, TGF-β is a potent inhibitor of proliferation. This growth factor acts by binding to a complex formed by TGF-β receptors I and II. Dimerization of receptor following ligand binding leads to a cascade of downstream signaling which results in transcriptional activation of growth inhibiting CDKIs and repression of growth promoting genes like MYC, CDK2, CDK4 and genes encoding cyclins A and E.

Mutations in TGF-β signaling pathway is common in pancreatic, colon, stomach and endometrial cancers. It is thought that TGF-β signaling pathway is involved in activation of epithelial to mesenchymal transition (EMT) in certain late stage malignant tumors.

Mutations Affecting Apoptosis

Abnormalities such as errors in mitosis and DNA damage can trigger apoptosis or programmed cell death. Apoptosis plays a very significant role in getting rid of excess and abnormal cells. If cells do not die when they should and instead keep proliferating, a tumor may form. For example:

- In chronic lymphoblastic leukemia (CLL), cells have chromosomal translocation that activate bcl-2 gene which is known to be a critical blocker of apoptosis. This inhibits apoptosis and allows survival of tumor cells.
- The most common tumor suppressor gene implicated in human cancers is p53. P53 induced expression of pro-apoptotic protein like Bax lead to programmed death of cells having extensive DNA damage. Cancers having mutated p53 gene lack proper apoptotic response and accumulation of abnormal and tumor cells.
- PTEN is another tumor suppressor gene implicated in several cancers. PTEN normally acts as a pro-apoptotic protein by dephosphorylating phosphatidylinositol 3,4,5-triphosphate, a secondary messenger that functions in activation of Akt pathway which promotes cell survival, growth and proliferation and prevents apoptosis by several pathways.

Molecular Composition of Cells

Cells are composed of water, inorganic ions, and carbon-containing (organic) molecules. Water is the most abundant molecule in cells, accounting for 70% or more of total cell mass. Consequently, the interactions between water and the other constituents of cells are of central importance in biological chemistry. The critical property of water in this respect is that it is a polar molecule, in which the hydrogen atoms have a slight positive charge and the oxygen has a slight negative charge. Because of their polar nature, water molecules can form hydrogen bonds with each other or with other polar molecules, as well as interacting with positively or negatively charged ions. As a result of these interactions, ions and polar molecules are readily soluble in water (hydrophilic). In contrast, nonpolar molecules, which cannot interact with water, are poorly soluble in an aqueous environment (hydrophobic). Consequently, nonpolar molecules tend to minimize their contact with water by associating closely with each other instead. Such interactions of polar and nonpolar molecules with water and with each other play crucial roles in the formation of biological structures, such as cell membranes.

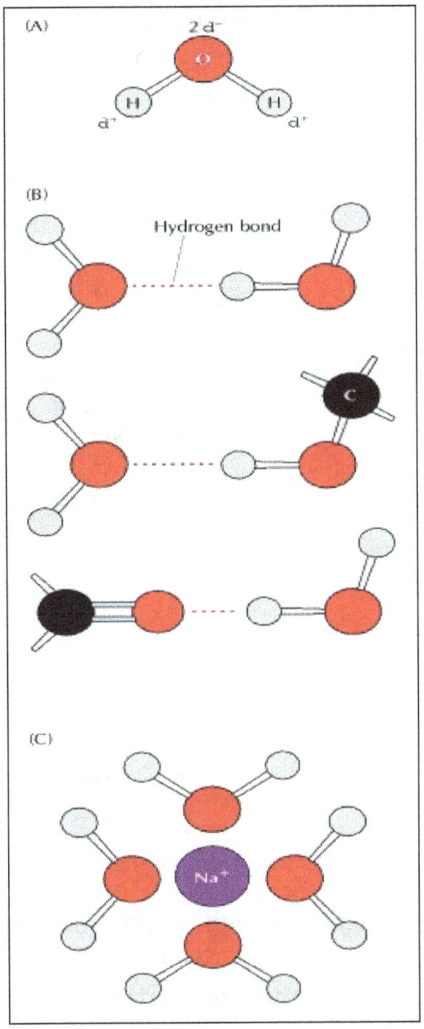

Figure shows characteristics of water. (A) Water is a polar molecule, with a slight negative charge

(δ^-) on the oxygen atom and a slight positive charge (δ^+) on the hydrogen atoms. Because of this polarity, water molecules can form hydrogen bonds (dashed lines) either with each other or with other polar molecules (B), in addition to interacting with charged ions (C).

The inorganic ions of the cell, including sodium (Na^+), potassium (K^+), magnesium (Mg^{2+}), calcium (Ca^{2+}), phosphate (HPO_4^{2-}), chloride (Cl^-), and bicarbonate (HCO_3^-), constitute 1% or less of the cell mass. These ions are involved in a number of aspects of cell metabolism, and thus play critical roles in cell function.

It is, however, the organic molecules that are the unique constituents of cells. Most of these organic compounds belong to one of four classes of molecules: carbohydrates, lipids, proteins, and nucleic acids. Proteins, nucleic acids, and most carbohydrates (the polysaccharides) are macromolecules formed by the joining (polymerization) of hundreds or thousands of low-molecular-weight precursors: amino acids, nucleotides, and simple sugars, respectively. Such macromolecules constitute 80 to 90% of the dry weight of most cells. Lipids are the other major constituent of cells. The remainder of the cell mass is composed of a variety of small organic molecules, including macromolecular precursors. The basic chemistry of cells can thus be understood in terms of the structures and functions of four major classes of organic molecules.

Carbohydrates

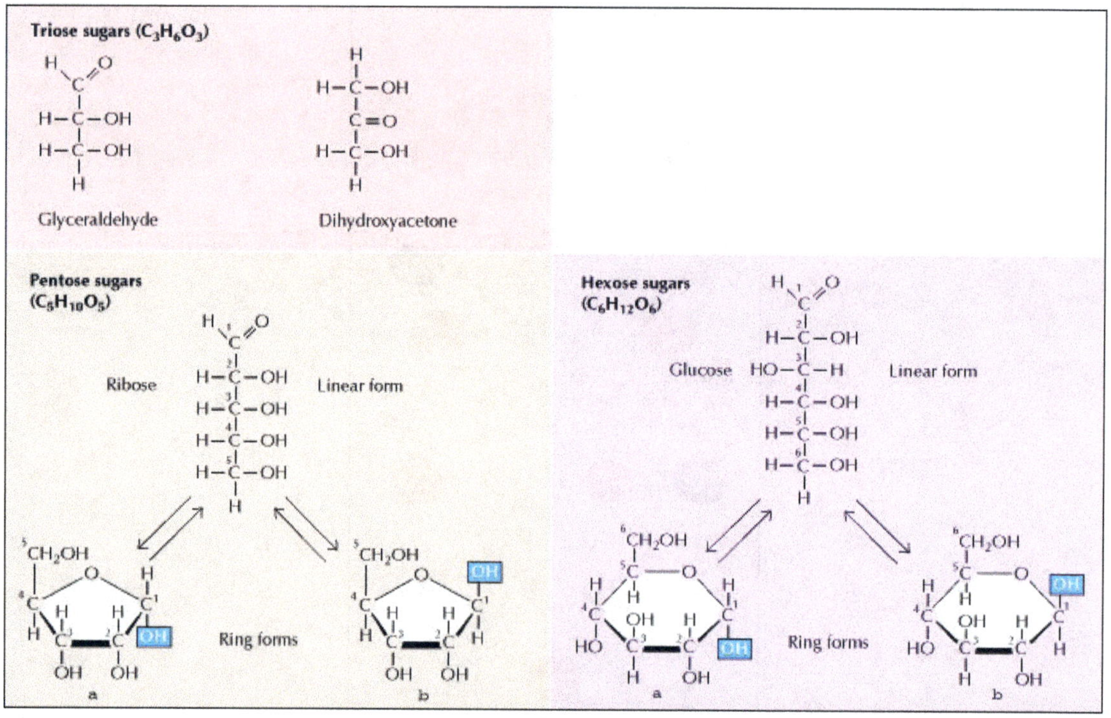

Structure of simple sugars.

The carbohydrates include simple sugars as well as polysaccharides. These simple sugars, such as glucose, are the major nutrients of cells. As discussed later in this chapter, their breakdown provides both a source of cellular energy and the starting material for the synthesis of other cell constituents. Polysaccharides are storage forms of sugars and form structural components of the cell. In addition, polysaccharides and shorter polymers of sugars act as markers for a variety of cell recognition processes, including the adhesion of cells to their neighbors and the transport of proteins to appropriate intracellular destinations.

The structures of representative simple sugars (monosaccharides) are illustrated in figure. The basic formula for these molecules is $(CH_2O)n$, from which the name carbohydrate is derived (C= "carbo" and H_2O= "hydrate"). The six-carbon (n= 6) sugar glucose ($C_6H_{12}O_6$) is especially important in cells, since it provides the principal source of cellular energy. Other simple sugars have between three and seven carbons, with three- and five-carbon sugars being the most common. Sugars containing five or more carbons can cyclize to form ring structures, which are the predominant forms of these molecules within cells. As illustrated in figure below, the cyclized sugars exist in two alternative forms (called α or β), depending on the configuration of carbon 1.

Representative sugars containing three, five, and six carbons (triose, pentose, and hexose sugars, respectively) are illustrated. Sugars with five or more carbons can cyclize to form rings, which exist in two alternative forms (α and β) depending on the configuration of carbon 1.

Monosaccharides can be joined together by dehydration reactions, in which H_2O is removed and the sugars are linked by a glycosidic bond between two of their carbons. If only a few sugars are joined together, the resulting polymer is called an oligosaccharide. If a large number (hundreds or thousands) of sugars are involved, the resulting polymers are macromolecules called polysaccharides.

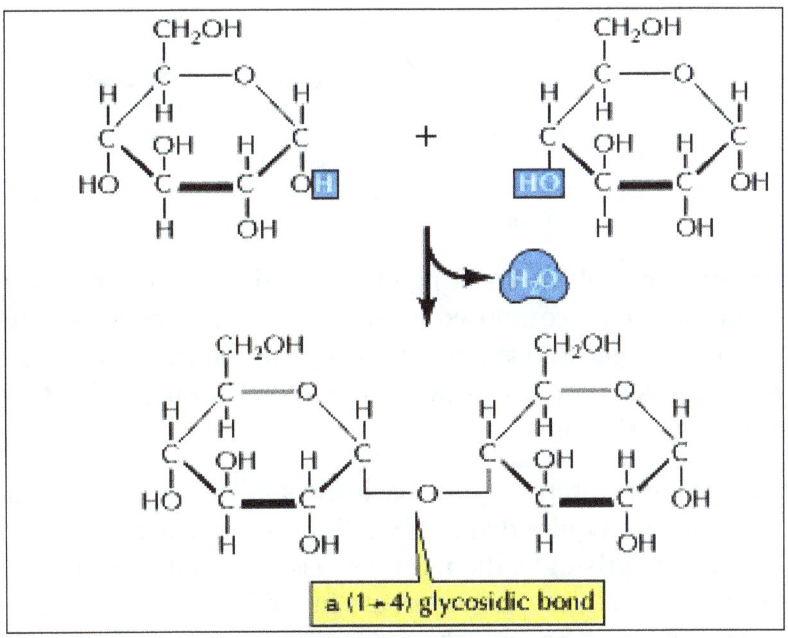

Formation of a glycosidic bond.

Two simple sugars are joined by a dehydration reaction (a reaction in which water is removed). In the example shown, two glucose molecules in the α configuration are joined by a bond between carbons 1 and 4, which is therefore called an α (1→4) glycosidic bond.

Two common polysaccharides—glycogen and starch—are the storage forms of carbohydrates in animal and plant cells, respectively. Both glycogen and starch are composed entirely of glucose molecules in the α configuration. The principal linkage is between carbon 1 of one glucose and carbon 4 of a second. In addition, both glycogen and one form of starch (amylopectin) contain occasional α (1→6) linkages, in which carbon 1 of one glucose is joined to carbon 6 of a second. As illustrated in Figure below, these linkages lead to the formation of branches resulting from the joining of two separate α (1→4) linked chains. Such branches are present in glycogen and amylopectin, although another form of starch (amylose) is an unbranched molecule.

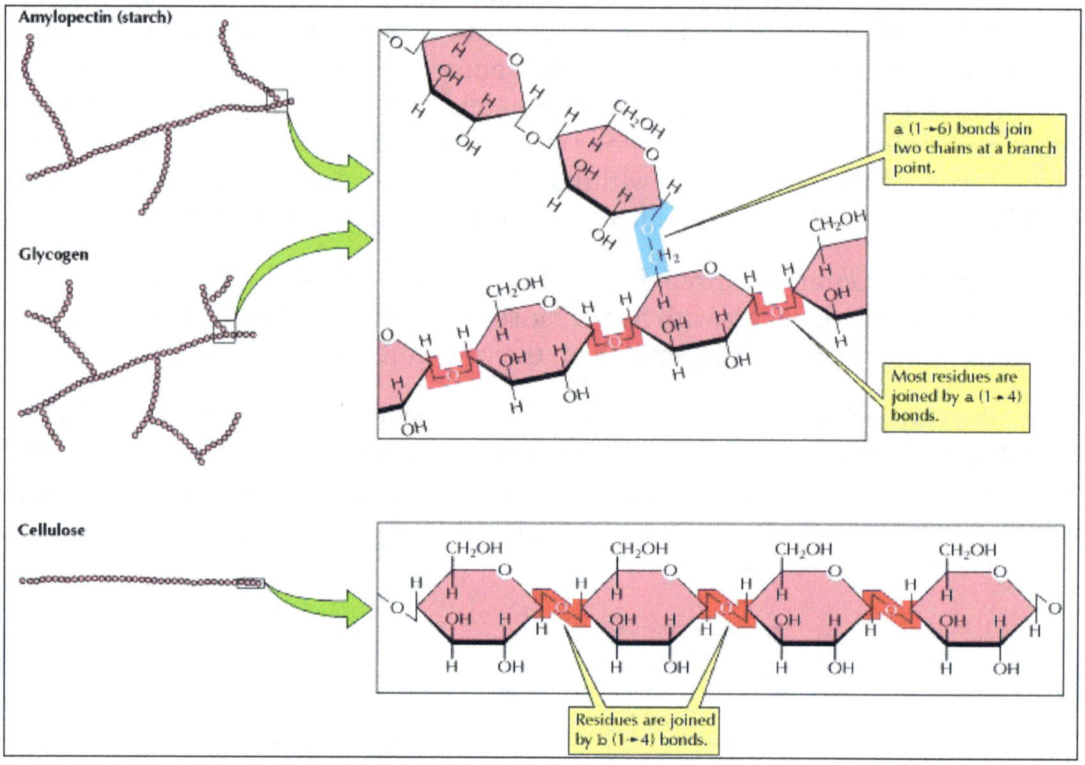

Structure of polysaccharides.

Polysaccharides are macromolecules consisting of hundreds or thousands of simple sugars. Glycogen, starch, and cellulose are all composed entirely of glucose residues, which are joined by α (1→4) glycosidic bonds in glycogen and starch but by β (1→4) bonds in cellulose. Glycogen and one form of starch (amylopectin) also contain occasional α (1→6) bonds, which serve as branch points by joining two separate α (1→4) chains.

The structures of glycogen and starch are thus basically similar, as is their function: to store glucose. Cellulose, in contrast, has a quite distinct function as the principal structural component of the plant cell wall. Perhaps surprisingly, then, cellulose is also composed entirely of glucose molecules. The glucose residues in cellulose, however, are in the β rather than the α configuration, and cellulose is an unbranched polysaccharide. The linkage of glucose residues by β (1→4) rather than α (1→4) bonds causes cellulose to form long extended chains that pack side by side to form fibers of great mechanical strength.

In addition to their roles in energy storage and cell structure, oligosaccharides and polysaccharides are important in a variety of cell signaling processes. For example, oligosaccharides are frequently linked to proteins, where they serve as markers to target proteins for transport to the cell surface or incorporation into different subcellular organelles. Oligosaccharides and polysaccharides also serve as markers on the surface of cells, playing important roles in cell recognition and the interactions between cells in tissues of multicellular organisms.

Lipids

Lipids have three major roles in cells. First, they provide an important form of energy storage. Second, and of great importance in cell biology, lipids are the major components of cell membranes. Third, lipids play important roles in cell signaling, both as steroid hormones (e.g., estrogen and testosterone) and as messenger molecules that convey signals from cell surface receptors to targets within the cell.

The simplest lipids are fatty acids, which consist of long hydrocarbon chains, most frequently containing 16 or 18 carbon atoms, with a carboxyl group (COO^-) at one end. Unsaturated fatty acids contain one or more double bonds between carbon atoms; in saturated fatty acids all of the carbon atoms are bonded to the maximum number of hydrogen atoms. The long hydrocarbon chains of fatty acids contain only nonpolar C—H bonds, which are unable to interact with water. The hydrophobic nature of these fatty acid chains is responsible for much of the behavior of complex lipids, particularly in the formation of biological membranes.

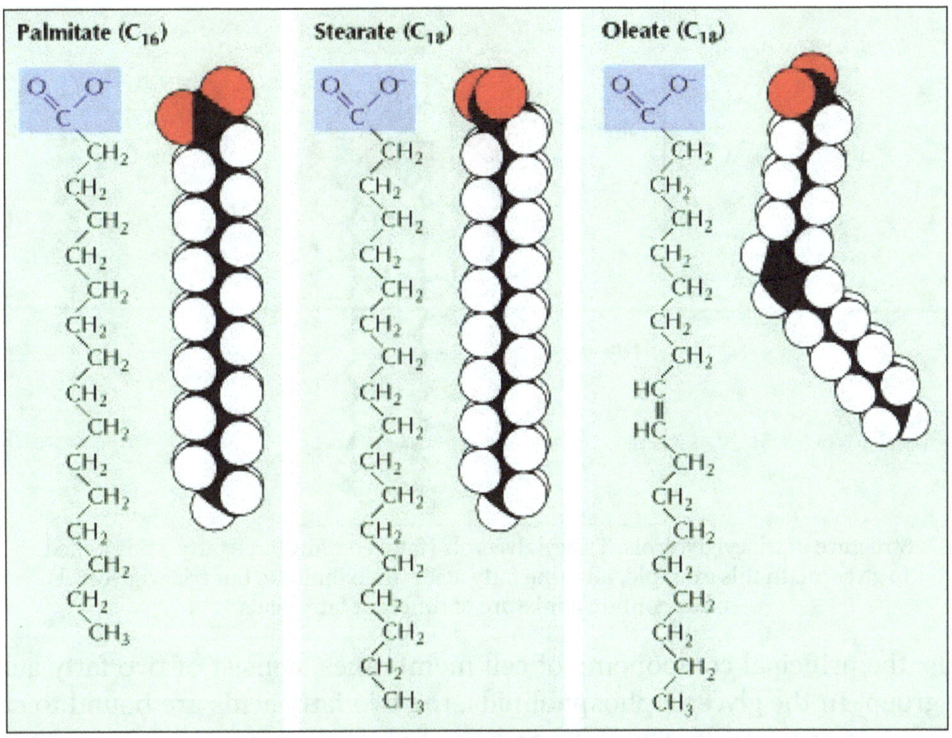

Structure of fatty acids.

Fatty acids consist of long hydrocarbon chains terminating in a carboxyl group (COO^-). Palmitate and stearate are saturated fatty acids consisting of 16 and 18 carbons, respectively. Oleate is an unsaturated 18-carbon fatty acid containing a double bond between carbons 9 and 10. Note that the double bond introduces a kink in the hydrocarbon chain.

Fatty acids are stored in the form of triacylglycerols, or fats, which consist of three fatty acids linked to a glycerol molecule. Triacylglycerols are insoluble in water and therefore accumulate as fat droplets in the cytoplasm. When required, they can be broken down for use in energy-yielding reactions discussed. It is noteworthy that fats are a more efficient form of energy storage than carbohydrates, yielding more than twice as much energy per weight of material broken down. Fats therefore allow energy to be stored in less than half the body weight that would be required to store the same amount of energy in carbohydrates—a particularly important consideration for animals because of their mobility.

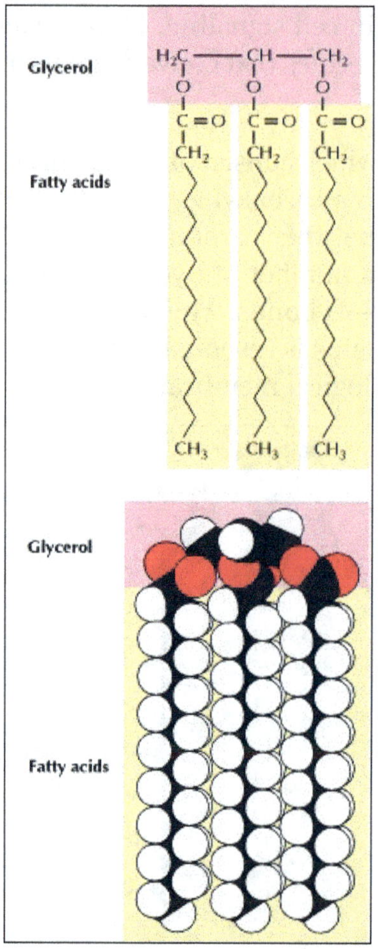

Structure of triacylglycerols. Triacylglycerols (fats) contain three fatty acids joined to glycerol. In this example, all three fatty acids are palmitate, but triacylglycerols often contain a mixture of different fatty acids.

Phospholipids, the principal components of cell membranes, consist of two fatty acids joined to a polar head group. In the glycerol phospholipids, the two fatty acids are bound to carbon atoms in glycerol, as in triacylglycerols. The third carbon of glycerol, however, is bound to a phosphate group, which is in turn frequently attached to another small polar molecule, such as choline, serine, inositol, or ethanolamine. Sphingomyelin, the only non-glycerol phospholipid in cell membranes, contains two hydrocarbon chains linked to a polar head group formed from serine rather than from glycerol. All phospholipids have hydrophobic tails, consisting of the two hydrocarbon chains, and hydrophilic head groups, consisting of the phosphate group and its polar attachments. Consequently, phospholipids are amphipathic molecules, part water-soluble and part water-insoluble. This property of phospholipids is the basis for the formation of biological membranes.

Glycerol phospholipids contain two fatty acids joined to glycerol. The fatty acids may be different from each other and are designated R1 and R2. The third carbon of glycerol is joined to a phosphate group (forming phosphatidic acid), which in turn is frequently joined to another small polar molecule (forming phosphatidylethanolamine, phosphatidylcholine, phosphatidylserine, or phosphatidylinositol). In sphingomyelin, two hydrocarbon chains are bound to a polar head group formed from serine instead of glycerol.

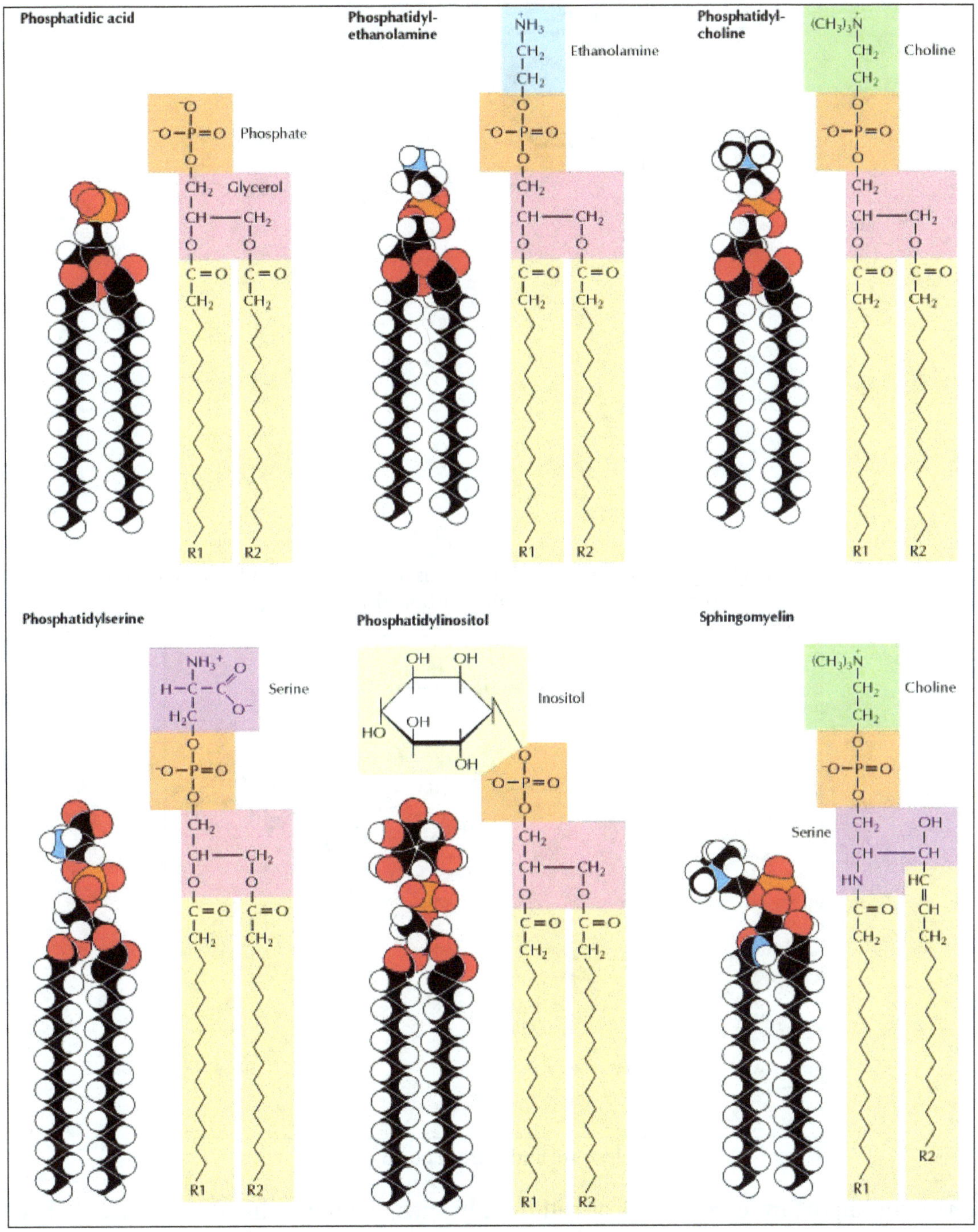

Structure of phospholipids.

In addition to phospholipids, many cell membranes contain glycolipids and cholesterol. Glycolipids consist of two hydrocarbon chains linked to polar head groups that contain carbohydrates. They are thus similar to the phospholipids in their general organization as amphipathic molecules.

Cholesterol, in contrast, consists of four hydrocarbon rings rather than linear hydrocarbon chains. The hydrocarbon rings are strongly hydrophobic, but the hydroxyl (OH) group attached to one end of cholesterol is weakly hydrophilic, so cholesterol is also amphipathic.

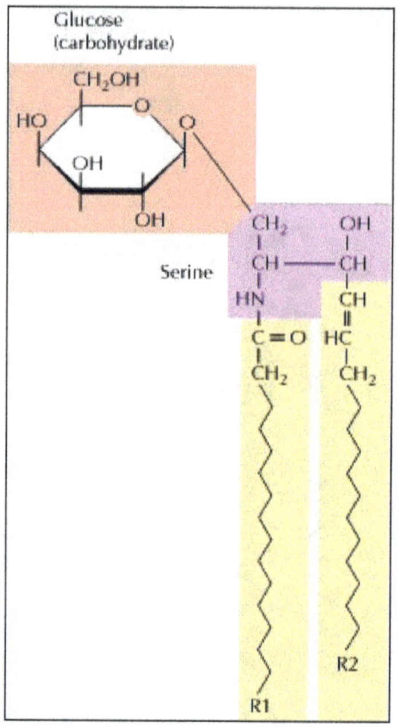

Structure of glycolipids. Two hydrocarbon chains are joined to a polar head group formed from serine and containing carbohydrates (e.g., glucose).

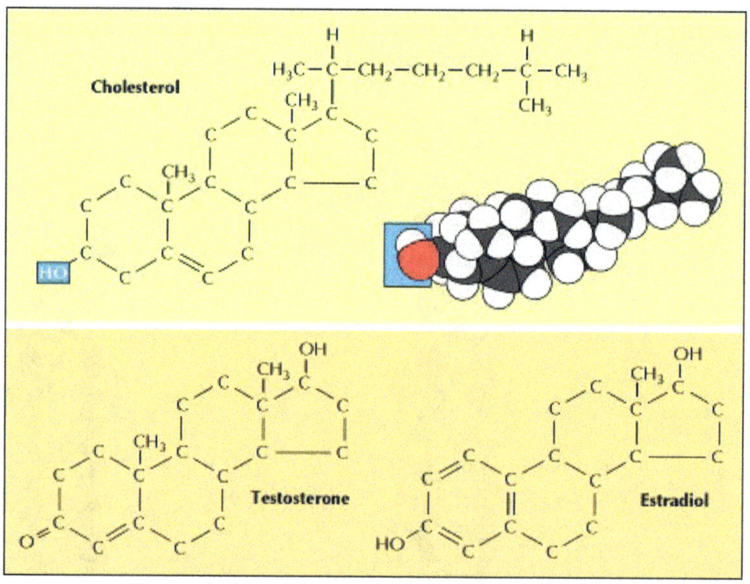

Cholesterol and steroid hormones.

Cholesterol, an important component of cell membranes, is an amphipathic molecule because of its polar hydroxyl group. Cholesterol is also a precursor to the steroid hormones, such as testosterone and estradiol (a form of estrogen). The hydrogen atoms bonded to the ring carbons are not shown in this figure.

In addition to their roles as components of cell membranes, lipids function as signaling molecules, both within and between cells. The steroid hormones (such as estrogens and testosterone) are derivatives of cholesterol. These hormones are a diverse group of chemical messengers, all of which contain four hydrocarbon rings to which distinct functional groups are attached. Derivatives of phospholipids also serve as messenger molecules within cells, acting to convey signals from cell surface receptors to intracellular targets.

Nucleic Acids

The nucleic acids—DNA and RNA—are the principal informational molecules of the cell. Deoxyribonucleic acid (DNA) has a unique role as the genetic material, which in eukaryotic cells is located in the nucleus. Different types of ribonucleic acid (RNA) participate in a number of cellular activities. Messenger RNA (mRNA) carries information from DNA to the ribosomes, where it serves as a template for protein synthesis. Two other types of RNA (ribosomal RNA and transfer RNA) are involved in protein synthesis. Still other kinds of RNAs are involved in the processing and transport of both RNAs and proteins. In addition to acting as an informational molecule, RNA is also capable of catalyzing a number of chemical reactions. In present-day cells, these include reactions involved in both protein synthesis and RNA processing.

DNA and RNA are polymers of nucleotides, which consist of purine and pyrimidine bases linked to phosphorylated sugars. DNA contains two purines (adenine and guanine) and two pyrimidines (cytosine and thymine). Adenine, guanine, and cytosine are also present in RNA, but RNA contains uracil in place of thymine. The bases are linked to sugars (2′-deoxyribose in DNA, or ribose in RNA) to form nucleosides. Nucleotides additionally contain one or more phosphate groups linked to the 5′ carbon of nucleoside sugars.

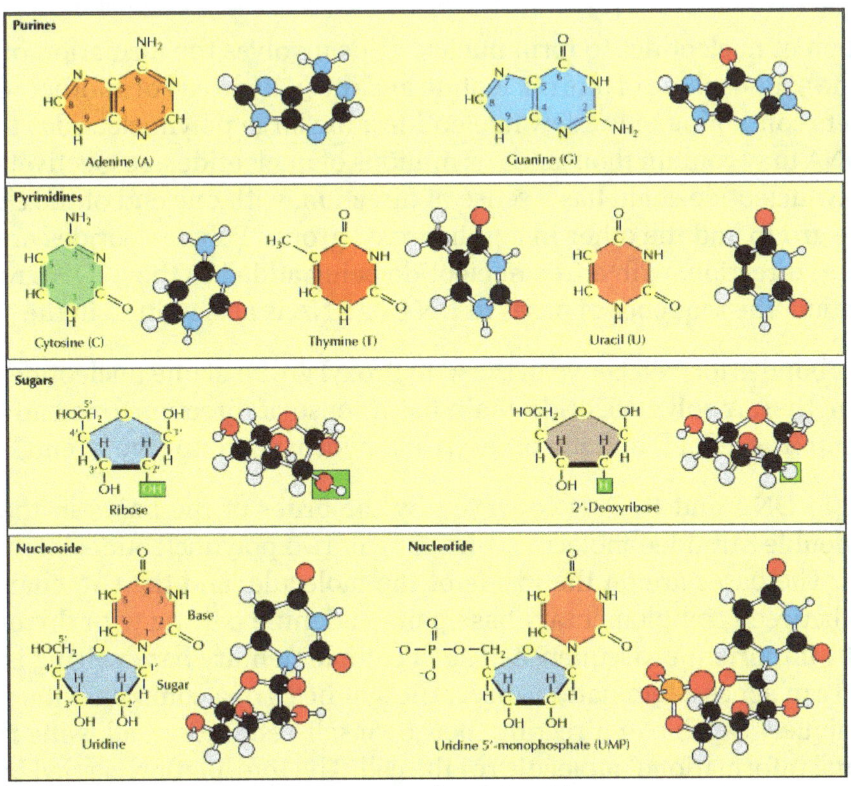

Components of nucleic acids.

Nucleic acids contain purine and pyrimidine bases linked to phosphorylated sugars. A nucleic acid base linked to a sugar alone is a nucleoside. Nucleotides additionally contain one or more phosphate groups.

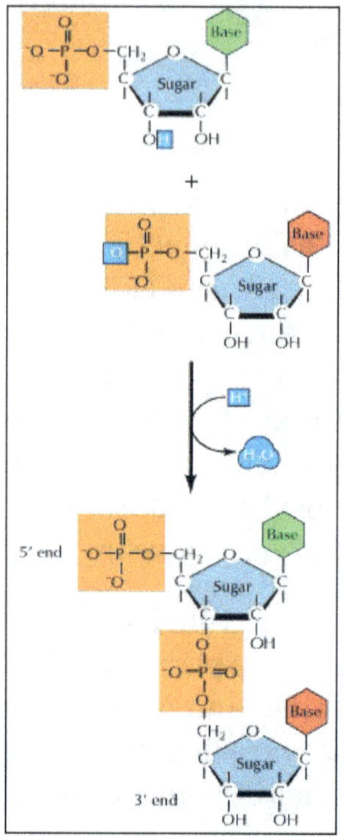

Polymerization of nucleotides.

The polymerization of nucleotides to form nucleic acids involves the formation of phosphodiester bonds between the 5′ phosphate of one nucleotide and the 3′ hydroxyl of another. Oligonucleotides are small polymers containing only a few nucleotides; the large polynucleotides that make up cellular RNA and DNA may contain thousands or millions of nucleotides, respectively. It is important to note that a polynucleotide chain has a sense of direction, with one end of the chain terminating in a 5′ phosphate group and the other in a 3′ hydroxyl group. Polynucleotides are always synthesized in the 5′ to 3′ direction, with a free nucleotide being added to the 3′ OH group of a growing chain. By convention, the sequence of bases in DNA or RNA is also written in the 5′ to 3′ direction.

A phosphodiester bond is formed between the 3′ hydroxyl group of one nucleotide and the 5′ phosphate group of another. A polynucleotide chain has a sense of direction, one end terminating in a 5′ phosphate group (the 5′ end) and the other in a 3′ hydroxyl group (the 3′ end).

The information in DNA and RNA is conveyed by the order of the bases in the polynucleotide chain. DNA is a double-stranded molecule consisting of two polynucleotide chains running in opposite directions. The bases are on the inside of the molecule, and the two chains are joined by hydrogen bonds between complementary base pairs—adenine pairing with thymine and guanine with cytosine. The important consequence of such complementary base pairing is that one strand of DNA (or RNA) can act as a template to direct the synthesis of a complementary strand. Nucleic acids are thus uniquely capable of directing their own self-replication, allowing them to function as the fundamental informational molecules of the cell. The information carried by DNA and RNA directs the synthesis of specific proteins, which control most cellular activities.

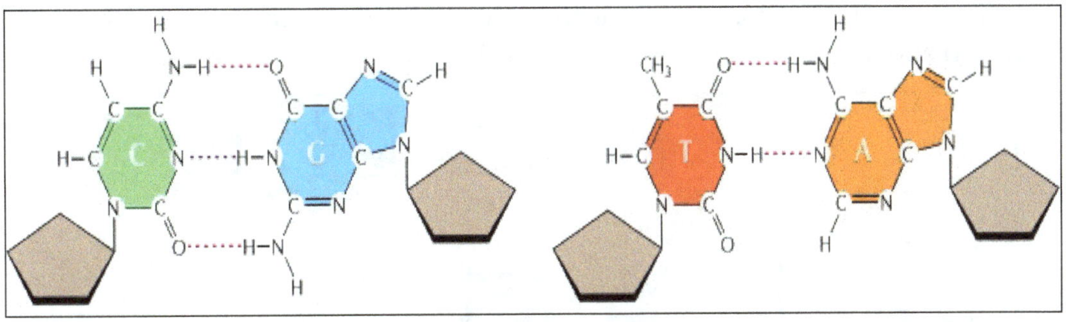

Complementary pairing between nucleic acid bases. The formation of hydrogen bonds between bases on opposite strands of DNA leads to the specific pairing of guanine (G) with cytosine (C) and adenine (A) with thymine (T).

Nucleotides are not only important as the building blocks of nucleic acids; they also play critical roles in other cell processes. Perhaps the most prominent example is adenosine 5′-triphosphate (ATP), which is the principal form of chemical energy within cells. Other nucleotides similarly function as carriers of either energy or reactive chemical groups in a wide variety of metabolic reactions. In addition, some nucleotides (e.g., cyclic AMP) are important signaling molecules within cells.

Proteins

While nucleic acids carry the genetic information of the cell, the primary responsibility of proteins is to execute the tasks directed by that information. Proteins are the most diverse of all macromolecules, and each cell contains several thousand different proteins, which perform a wide variety of functions. The roles of proteins include serving as structural components of cells and tissues, acting in the transport and storage of small molecules (e.g., the transport of oxygen by hemoglobin), transmitting information between cells (e.g., protein hormones), and providing a defense against infection (e.g., antibodies). The most fundamental property of proteins, however, is their ability to act as enzymes, which, catalyze nearly all the chemical reactions in biological systems. Thus, proteins direct virtually all activities of the cell. The central importance of proteins in biological chemistry is indicated by their name, which is derived from the Greek word proteios, meaning "of the first rank."

Proteins are polymers of 20 different amino acids. Each amino acid consists of a carbon atom (called the α carbon) bonded to a carboxyl group (COO^-), an amino group (NH_3^+), a hydrogen atom, and a distinctive side chain. The specific chemical properties of the different amino acid side chains determine the roles of each amino acid in protein structure and function.

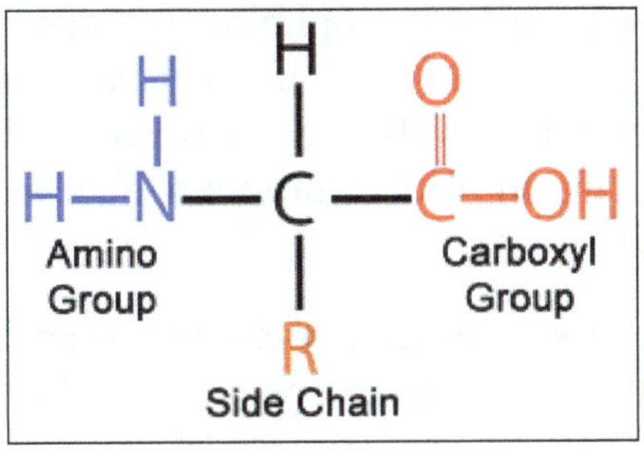

Structure of amino acids.

Each amino acid consists of a central carbon atom (the α carbon) bonded to a hydrogen atom, a carboxyl group, an amino group, and a specific side chain (designated R). At physiological pH, both the carboxyl and amino groups are ionized, as shown.

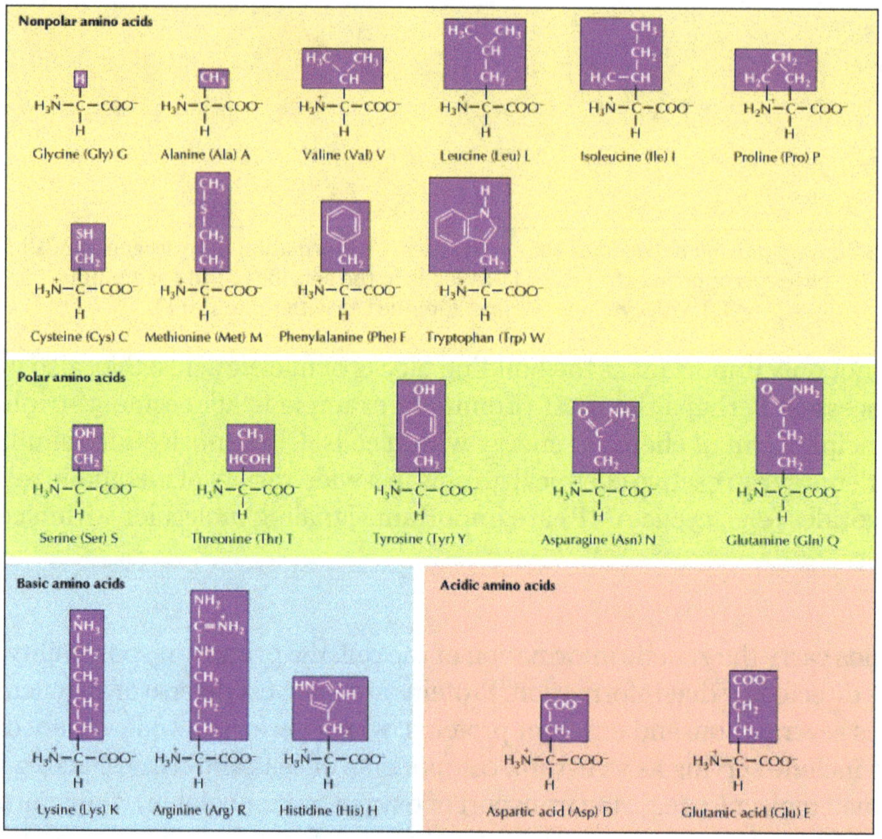

The amino acids.

The amino acids can be grouped into four broad categories according to the properties of their side chains. Ten amino acids have nonpolar side chains that do not interact with water. Glycine is the simplest amino acid, with a side chain consisting of only a hydrogen atom. Alanine, valine, leucine, and isoleucine have hydrocarbon side chains consisting of up to four carbon atoms. The side chains of these amino acids are hydrophobic and therefore tend to be located in the interior of proteins, where they are not in contact with water. Proline similarly has a hydrocarbon side chain, but it is unique in that its side chain is bonded to the nitrogen of the amino group as well as to the α carbon, forming a cyclic structure. The side chains of two amino acids, cysteine and methionine, contain sulfur atoms. Methionine is quite hydrophobic, but cysteine is less so because of its sulfhydryl (SH) group. The sulfhydryl group of cysteine plays an important role in protein structure because disulfide bonds can form between the side chains of different cysteine residues. Finally, two nonpolar amino acids, phenylalanine and tryptophan, have side chains containing very hydrophobic aromatic rings.

The three-letter and one-letter abbreviations for each amino acid are indicated. The amino acids are grouped into four categories according to the properties of their side chains: nonpolar, polar, basic, and acidic.

Five amino acids have uncharged but polar side chains. These include serine, threonine, and tyrosine, which have hydroxyl groups on their side chains, as well as asparagine and glutamine, which have polar amide (O=C—NH_2) groups. Because the polar side chains of these amino acids can form hydrogen bonds with water, these amino acids are hydrophilic and tend to be located on the outside of proteins.

The amino acids lysine, arginine, and histidine have side chains with charged basic groups. Lysine and arginine are very basic amino acids, and their side chains are positively charged in the cell. Consequently, they are very hydrophilic and are found in contact with water on the surface of proteins. Histidine can be either uncharged or positively charged at physiological pH, so it frequently plays an active role in enzymatic reactions involving the exchange of hydrogen ions, as illustrated in the example of enzymatic catalysis.

Finally, two amino acids, aspartic acid and glutamic acid, have acidic side chains terminating in carboxyl groups. These amino acids are negatively charged within the cell and are therefore frequently referred to as aspartate and glutamate. Like the basic amino acids, these acidic amino acids are very hydrophilic and are usually located on the surface of proteins.

Amino acids are joined together by peptide bonds between the α amino group of one amino acid and the α carboxyl group of a second. Polypeptides are linear chains of amino acids, usually hundreds or thousands of amino acids in length. Each polypeptide chain has two distinct ends, one terminating in an α amino group (the amino, or N, terminus) and the other in an α carboxyl group (the carboxy, or C, terminus). Polypeptides are synthesized from the amino to the carboxy terminus, and the sequence of amino acids in a polypeptide is written (by convention) in the same order.

The defining characteristic of proteins is that they are polypeptides with specific amino acid sequences. In 1953 Frederick Sanger was the first to determine the complete amino acid sequence of a protein, the hormone insulin. Insulin was found to consist of two polypeptide chains, joined by disulfide bonds between cysteine residues. Most important, Sanger's experiment revealed that each protein consists of a specific amino acid sequence. Proteins are currently sequenced using automated methods, and the complete amino acid sequences of over 100,000 proteins are now known. Each consists of a unique sequence of amino acids, determined by the order of nucleotides in a gene.

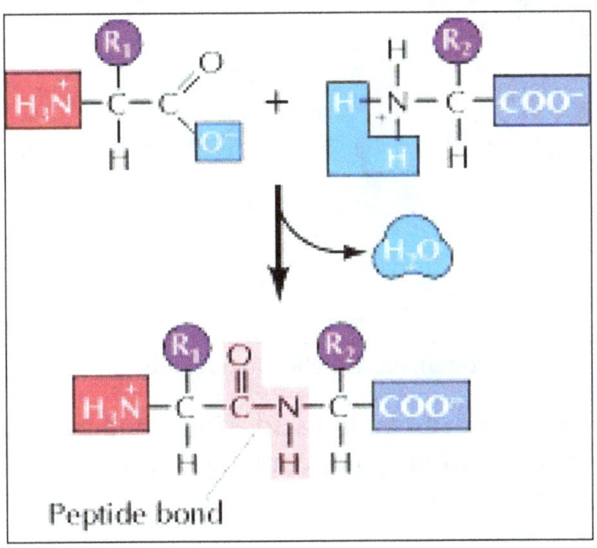

Formation of a peptide bond. The carboxyl group of one amino acid is linked to the amino group of a second.

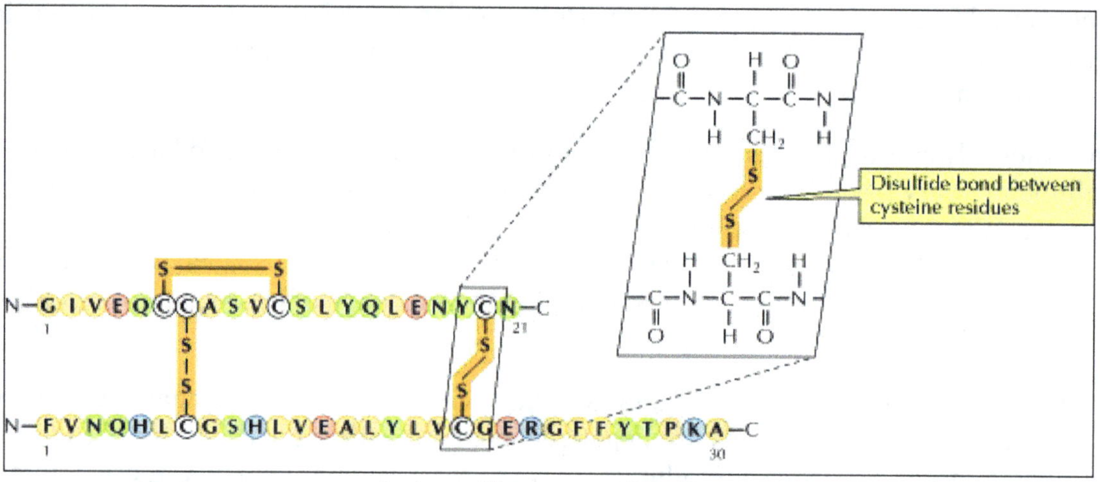

Amino acid sequence of insulin.

Insulin consists of two polypeptide chains, one of 21 and the other of 30 amino acids (indicated here by their one-letter codes). The side chains of three pairs of cysteine residues are joined by disulfide bonds, two of which connect the polypeptide chains.

The amino acid sequence of a protein is only the first element of its structure. Rather than being extended chains of amino acids, proteins adopt distinct three-dimensional conformations that are critical to their function. These three-dimensional conformations of proteins are the result of interactions between their constituent amino acids, so the shapes of proteins are determined by their amino acid sequences. This was first demonstrated by experiments of Christian Anfinsen in which he disrupted the three-dimensional structures of proteins by treatments, such as heating, that break non-covalent bonds—a process called denaturation. Following incubation under milder conditions, such denatured proteins often spontaneously returned to their native conformations, indicating that these conformations were directly determined by the amino acid sequence.

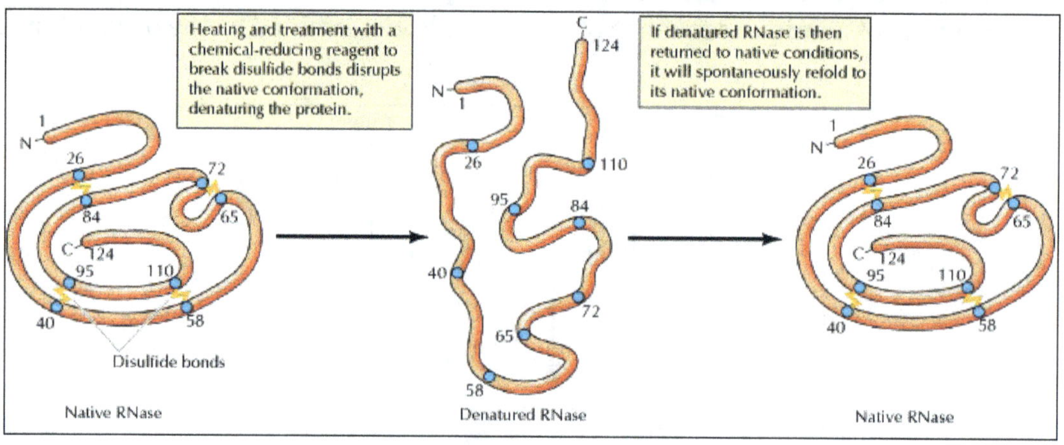

Protein denaturation and refolding.

Ribonuclease (RNase) is a protein of 124 amino acids (indicated by numbers). The protein is normally folded into its native conformation, which contains four disulfide bonds (indicated as paired circles representing the cysteine residues).

The three-dimensional structure of proteins is most frequently analyzed by X-ray crystallography, a high-resolution technique that can determine the arrangement of individual atoms within a molecule. A beam of X rays is directed at crystals of the protein to be analyzed, and the pattern of X rays that pass through the protein crystal is detected on X-ray film. As the X rays strike the crystal, they are scattered in characteristic patterns determined by the arrangement of atoms in the molecule. The structure of the molecule can therefore be deduced from the pattern of scattered X rays (the diffraction pattern).

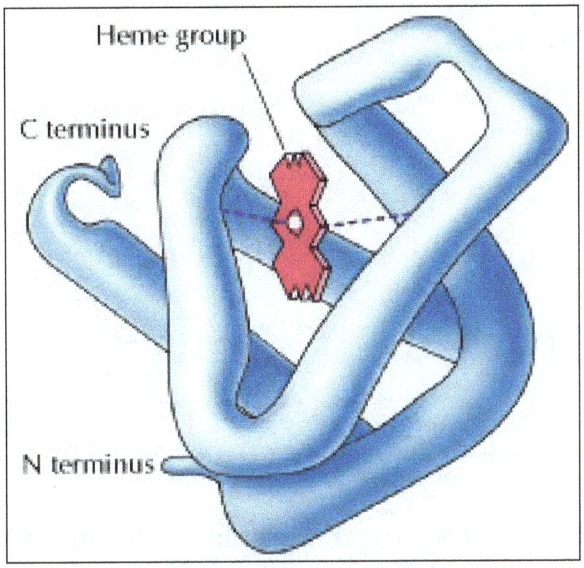

Three-dimensional structure of myoglobin.

In 1958 John Kendrew was the first to determine the three-dimensional structure of a protein, myoglobin—a relatively simple protein of 153 amino acids. Since then, the three-dimensional structures of several thousand proteins have been analyzed. Most, like myoglobin, are globular proteins with polypeptide chains folded into compact structures, although some (such as the structural proteins of connective tissues) are long fibrous molecules. Analysis of the three-dimensional structures of these proteins has revealed several basic principles that govern protein folding, although protein structure is so complex that predicting the three-dimensional structure of a protein directly from its amino acid sequence is impossible.

Myoglobin is a protein of 153 amino acids that is involved in oxygen transport. The polypeptide chain is folded around a heme group that serves as the oxygen-binding site.

Protein structure is generally described as having four levels. The primary structure of a protein is the sequence of amino acids in its polypeptide chain. The secondary structure is the regular arrangement of amino acids within localized regions of the polypeptide. Two types of secondary structure, which were first proposed by Linus Pauling and Robert Corey in 1951, are particularly common: the α helix and the β sheet. Both of these secondary structures are held together by hydrogen bonds between the CO and NH groups of peptide bonds. An α helix is formed when a region of a polypeptide chain coils around itself, with the CO group of one peptide bond forming a hydrogen bond with the NH group of a peptide bond located four residues downstream in the linear polypeptide chain. In contrast, a β sheet is formed when two parts of a polypeptide chain lie side by side with hydrogen bonds between them. Such β sheets can be formed between several polypeptide strands, which can be oriented either parallel or antiparallel to each other.

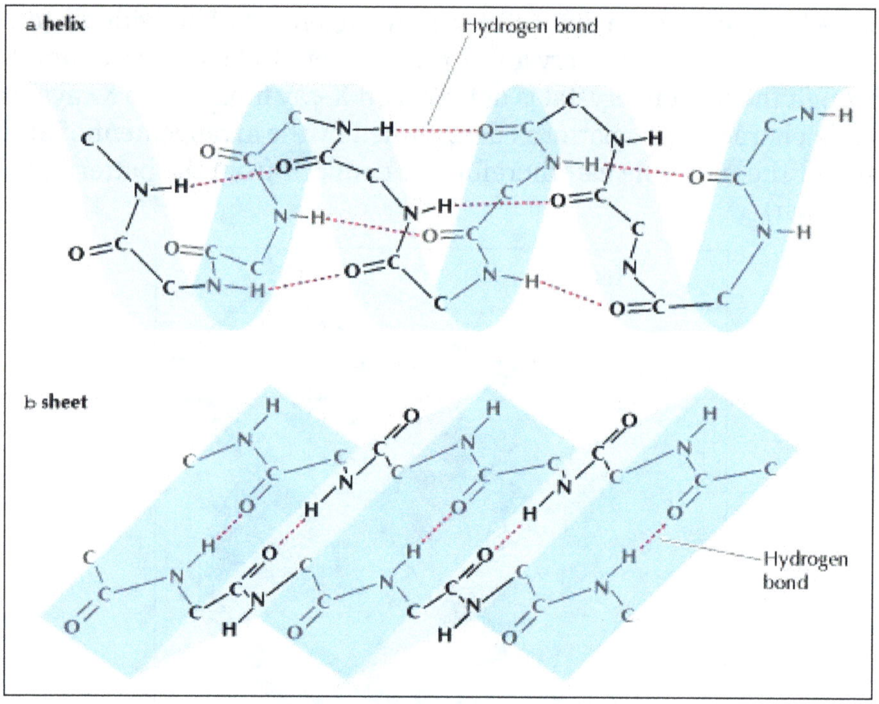

Secondary structure of proteins.

The most common types of secondary structure are the α helix and the β sheet. In an α helix, hydrogen bonds form between CO and NH groups of peptide bonds separated by four amino acid residues. In a β sheet, hydrogen bonds connect two parts of a polypeptide chain lying side by side. The amino acid side chains are not shown.

Tertiary structure is the folding of the polypeptide chain as a result of interactions between the side chains of amino acids that lie in different regions of the primary sequence. In most proteins, combinations of α helices and β sheets, connected by loop regions of the polypeptide chain, fold into compact globular structures called domains, which are the basic units of tertiary structure. Small proteins, such as ribonuclease or myoglobin, contain only a single domain; larger proteins may contain a number of different domains, which are frequently associated with distinct functions.

Regions of α-helix and β-sheet secondary structures, connected by loop regions, are folded into the native conformation of the protein. In this schematic representation of the polypeptide chain as a ribbon model, α helices are represented as spirals and β sheets as wide arrows.

A critical determinant of tertiary structure is the localization of hydrophobic amino acids in the interior of the protein and of hydrophilic amino acids on the surface, where they interact with water. The interiors of folded proteins thus consist mainly of hydrophobic amino acids arranged in α helices and β sheets; these secondary structures are found in the hydrophobic cores of proteins because hydrogen bonding neutralizes the polar character of the CO and NH groups of the polypeptide backbone. The loop regions connecting the elements of secondary structure are found on the surface of folded proteins, where the polar components of the peptide bonds form hydrogen bonds with water or with the polar side chains of hydrophilic amino acids. Interactions between polar amino acid side chains (hydrogen bonds and ionic bonds) on the protein surface are also important determinants of tertiary structure. In addition, the covalent disulfide bonds between the sulfhydryl groups of cysteine residues stabilize the folded structures of many cell-surface or secreted proteins.

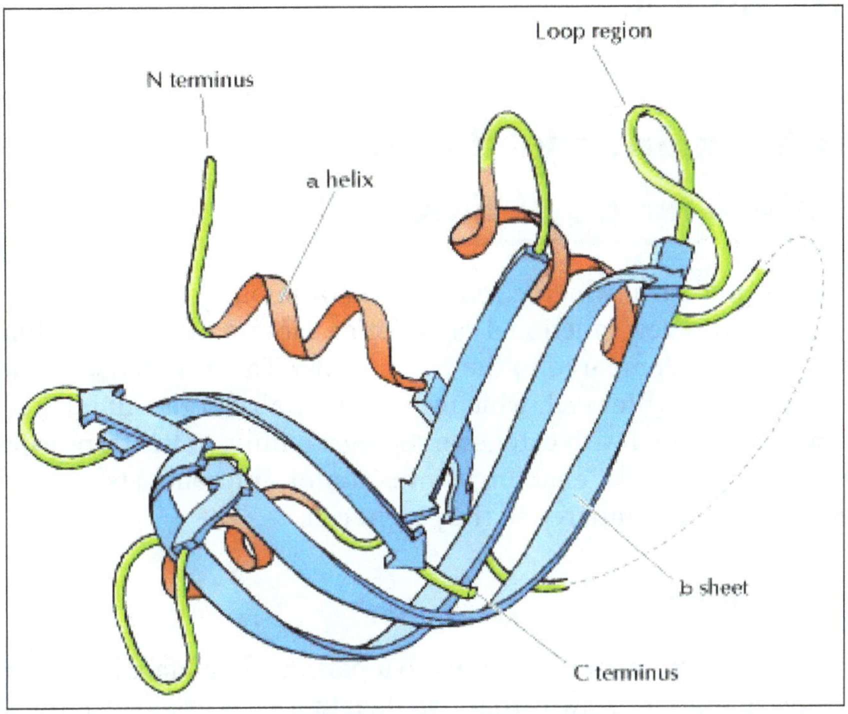

Tertiary structure of ribonuclease.

The fourth level of protein structure, quaternary structure, consists of the interactions between different polypeptide chains in proteins composed of more than one polypeptide. Hemoglobin, for example, is composed of four polypeptide chains held together by the same types of interactions that maintain tertiary structure.

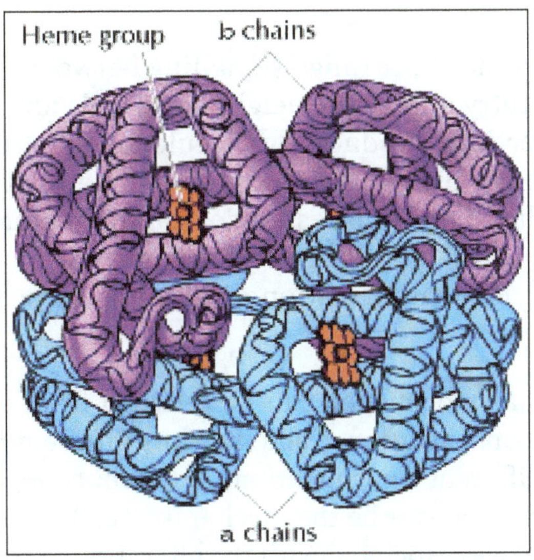

Quaternary structure of hemoglobin. Hemoglobin is composed of four poly-peptide chains, each of which is bound to a heme group. The two α chains and the two β chains are identical.

The distinct chemical characteristics of the 20 different amino acids thus lead to considerable variation in the three-dimensional conformations of folded proteins. Consequently, proteins constitute an extremely complex and diverse group of macromolecules, suited to the wide variety of tasks they perform in cell biology.

Molecular Mechanisms of Cell Differentiation and Cell Fates

Cell differentiation is a process in which a cell changes from one form to another. Differentiation usually provides the cell with new functional capabilities. How are these novel functions bestowed upon the differentiated cells, compared to the parent cells? This is because of the expression of a new set of proteins in the daughter cell, which were either absent or minimal in its predecessor. Thus differentiation is associated with extensive reprogramming at the transcriptional, post transcriptional and translational levels, enabling expression of different sets of proteins at different levels in the differentiated cell compared to the parent cell.

Cell Fate Determination

The process in which a cell irreversibly commits to a particular cell fate is known as cell determination or cell fate determination. How is that two daughter cells arising from a single parent cell can have different fates? Multiple mechanisms have been postulated to explain this. Cell division can result in asymmetrical distribution of molecules among the daughter cells- for instance, the mRNA content of the daughter cells can considerably vary from one another. If these mRNAs encode transcription factors, this can lead to altered transcriptional regulation and thus protein levels in the daughter cells. In other cases, it can be the transcription factors themselves that are differentially distributed.

Inductive signalling refers to the short range signalling between cells where signals from one set of cells of a different ancestry affects the neighboring cells and causes them to adapt a particular fate. Here the inducer cells produce a molecule, usually a protein, that interacts with the responder cell and leads to alteration in properties like mitotic rate, cell shape etc. The interaction between the inducer and the responder can be mediated by diffusible signals or direct cell to cell interaction or via gap junctions. Inductive signalling is particularly important during organ formation. Pattern formation deals with the positional orientation of a cell within an organ.

A diffusible chemical known as morphogen can be released from certain cells and this usually results in the formation of a concentration gradient. Thus different cells along the vicinity of the morphogen gets exposed to different concentrations of the morphogen and this leads to differential programming of the cells. Whatever be the mechanism, cell fate determination is followed by cellular differentiation and this is associated with the alteration in the protein types and amounts in the cells.

Mechanisms of Altered Protein Expression during Cellular Differentiation

Various mechanisms exist to regulation the expression of proteins in a cell. The processes important in cellular differentiation include, but are not restricted, to the following:

- Translational control: Processing bodies (P-bodies) are foci present in cytoplasm of eukaryotic cells that are associated with mRNA turn-over. Not all the mRNAs that enter the P-bodies are degraded. mRNAs can exit from the P-bodies and can further undergo translation. Thus certain mRNAs stored within the P-bodies are temporarily inaccessible to the translational machinery. This means that such proteins can be temporally expressed, depending on the cells involved.
- Gene amplification: Protein levels can be increased due to the increase in the number of gene encoding the protein, via selective amplification of the gene involved. Gene amplification is also seen in certain tumor tissues and has been one of the reasons for chemoresistance. Example: Multiple drug resistant (MDR) protein is involved in extrusion of certain drugs from the cells. Amplification of MDR proteins in certain cancers promote resistance against chemotherapeutic agents like methotrexate, which is actively extruded.
- Gene rearrangement: This is observed in B and T lymphocytes. The gene rearrangement in the B cell receptor and T cell receptor gene complexes, respectively, in these cells leads to formation of B and T cell receptors respectively. The specific pattern of rearrangement leads to diversity in the B and T cell receptors, thus enabling the immune system to respond to a wide variety of antigens. Defect in B and T cell receptor rearrangements can lead to immune deficiency.

Transcriptional Control

Expression of different sets of proteins is mainly regulated at the transcriptional level. Transcription factors can bind to the DNA at the promoter and enhancer regions and can promote the recruitment of RNA polymerases. These results in increased transcription thus increased mRNA levels and thus protein levels. Repressors are molecules that antagonize the action of inducers. They can bind to promoters or to DNA regions known as silencers.

Transcriptional control is achieved mostly via epigenetic mechanisms - cell differentiation is an epigenetic process. Epigenetics refers to the heritable changes in the genome without any change in the sequence of DNA. Epigenetic mechanisms play a prime role in regulation of gene expression. DNA methylation and histone modifications are the major epigenetic mechanisms.

5-methyl cytosine (5-mC), an epigenetic modification, mostly exists in the CpG dinucleotide context in the DNA -70-80% of CpGs are methylated. 5-mC is known as the 5th base of DNA. These are commonly found in the 5' end of the genes. The methyl group of 5-mC protrudes into the major groove of the DNA where it interferes with the binding of transcription factors. Thus they reduce transcription. CpG methylation is mediated by DNA methyl transferases (DNMTs) and the removal of this signature is mediated by enzymes like Ten Eleven Translocases (TET), which converts 5-mC to 5-hydroxymethyl cytosine (5-hmC, which is known as the 6th base of DNA) Eukaryotic DNA is restricted to the transcription factors. This is because of the compaction and the packing of the DNA which is mediated by histone proteins. Histones are positively charged.

They interact with the negatively charged phospho-diester backbone of the DNA and results in DNA condensation. One of the epigenetic modification is acetylation of lysine residues of histones. Acetylation results in loss of positive charge from lysine. Thus, a relatively less positively charged histone will have relatively less affinity to the negatively charged phospho-diesters of DNA. This leads to unwinding of the DNA making it more accessible to transcription factors. Two classes of enzymes- Histone Acetyl Transferases (HATs) and Histone Deacetylases (HDACs) are involved in this epigenetic modification.

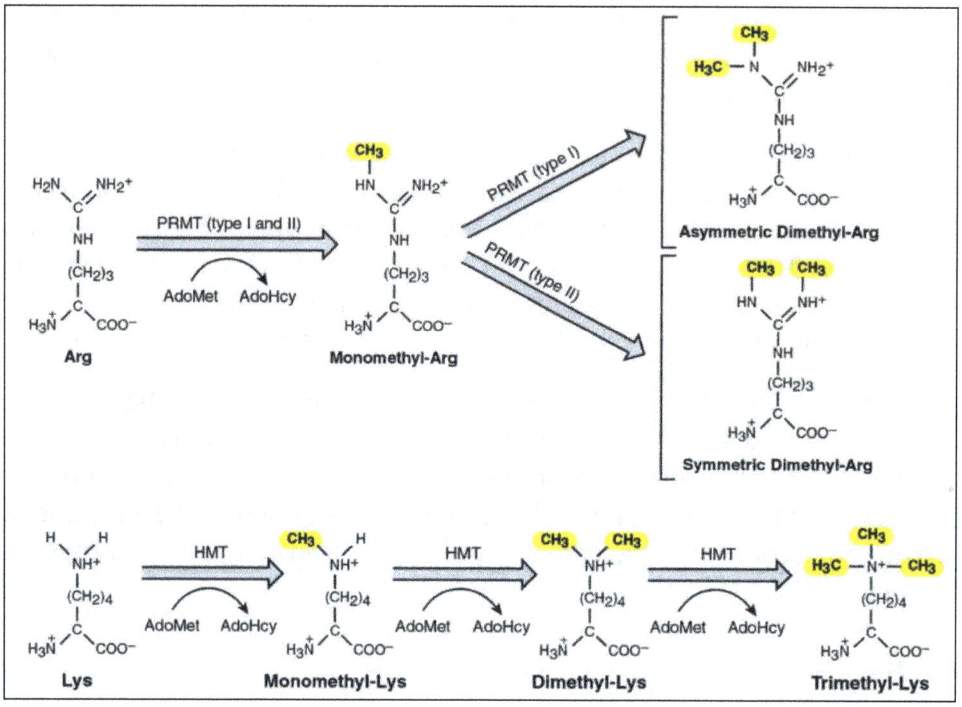

The other major histone modification is the methylation of amino acid residues of histones. Lysine and less commonly, arginine are the amino acid residues that are subjected to methylation. Unlike acetylation of lysine residues, histone methylation can increase or decrease transcription based on the position of the residue that is methylated, and the number of methyl groups. Lysine residues can be mono-, di-, or tri-methylated while arginine residues can be mono- or di-methylated. Addition of methyl groups to lysine is catalysed by Histone Methyl Transferases, while the similar reaction involving arginine is catalysed by Protein Arginine Methyl Transferases (PRMTs, R is the single letter notation for Arginine). S-adenosyl methionine (AdoMet) acts as the methyl donor in these reactions.

How exactly does histone methylations alter transcription? There are multiple mechanisms involved, but the prominently identified mechanism is that the methylated amino-acid residues can alter protein binding to the histones. Thus methylation can control recruitment of transcription factors to histones. The other class of proteins that are differentially recruited in response to methylation are the histone deacetylases.

Clinical Significance

Aberrant epigenetic mechanisms are associated with arrest of differentiation. This, if happens in hematopoietic stem cells can lead to abnormality in the formation of mature blood cells like RBCs, WBCs and platelets- thus causing anemia, leucopenia and infections, and thrombocytopenia and bleeding. This is what happens in myelodysplastic syndromes (MDS) which is associated with mutations in TET or DNMTs. Hypomethylating agents like Decitabine are used in the treatment of such conditions. HDAC inhibitor like suberoylanilide hydroxamic acid is used in the treatment of cutaneous T-cell lymphoma.

Is Cell Differentiation Reversible?

Let us consider 3 scenarios:

- Somatic Cell Nuclear Transfer (SCNT): In SCNT, the nucleus of a somatic cell is transferred to the cytoplasm of an enucleated cell (eg. ovum). Various factors present in the cytoplasm now leads to reprogramming of the somatic cell nucleus, conferring pluripotency, leading to generation of embryonic stem cells (ESCs). These cells can reach the blastocyst stage, when they can be used to isolate ESCs, or the blastocyst can be introduced into the uterus of a surrogate mother (reproductive cloning- e.g. Dolly).
- Induced Pluripotency: The introduction of selected transcription factors (Sox2, Oct4, Klf4 and cMyc) is sufficient to induce pluripotency in differentiated cells. Studies have shown that these transcription factors can lead to alterations in DNA methylation patterns and histone modifications, thus enabling the expression of genes that regulate pluripotency.
- Direct Reprogramming: This refers to the ability of an adult cell type to become another adult cell type without passing through a pluripotent state. The classical example of direct reprogramming is the conversion of fibroblasts to skeletal muscle cells by ectopic expression of a single gene, myogenic differentiation 1 (MYOD1).

All these examples of reversible differentiation or reprogramming accentuate the fact that differential protein expression at different phases of cell development leads to cellular differentiation.

Epigenetics

Epigenetics is the study of the chemical modification of specific genes or gene-associated proteins of an organism. Epigenetic modifications can define how the information in genes is expressed and used by cells. The term epigenetics came into general use in the early 1940s, when British embryologist Conrad Waddington used it to describe the interactions between genes and gene products, which direct development and give rise to an organism's phenotype (observable characteristics). Since then, information revealed by epigenetics studies has revolutionized the fields of genetics and developmental biology. Specifically, researchers have uncovered a range of possible chemical modifications to deoxyribonucleic acid (DNA) and to proteins called histones that associate tightly with DNA in the nucleus. These modifications can determine when or even if a given gene is expressed in a cell or an organism.

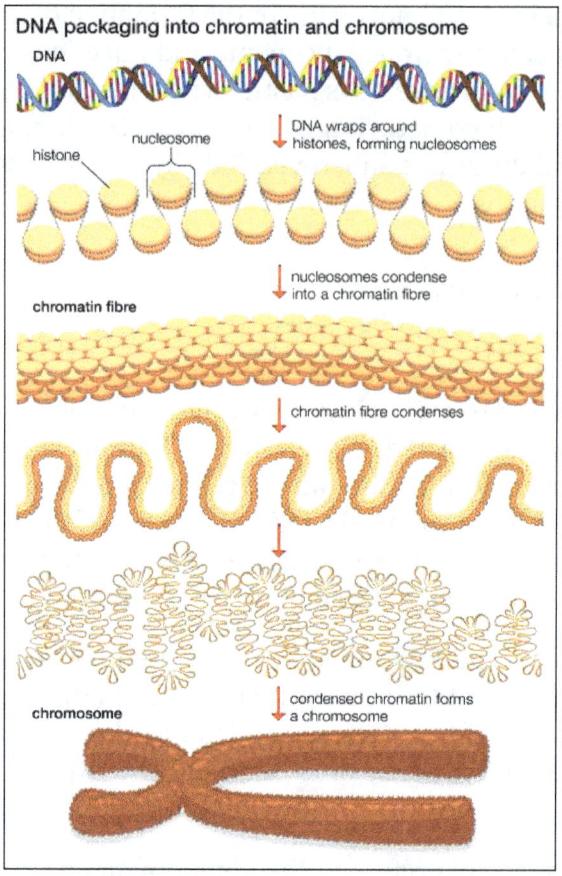

DNA wraps around proteins called histones to form units known as nucleosomes. These units condense into a chromatin fibre, which condenses further to form a chromosome. Epigenetics studies have revealed that chemical modifications to histones can be inherited and define how the information in genes is expressed and used by cells.

Types of Epigenetic Modifications

The principal type of epigenetic modification that is understood is methylation (addition of a methyl group). Methylation can be transient and can change rapidly during the life span of a cell or organism, or it can be essentially permanent once set early in the development of the embryo. Other largely permanent chemical modifications also play a role; these include histone acetylation (addition of an acetyl group), ubiquitination (the addition of a ubiquitin protein), and phosphorylation (the addition of a phosphoryl group). The specific location of a given chemical modification can also be important. For example, certain histone modifications distinguish actively expressed regions of the genome from regions that are not highly expressed. These modifications may correlate with chromosome banding patterns generated by staining procedures common in karyotype analyses. Similarly, specific histone modifications may distinguish actively expressed genes from genes that are poised for expression or genes that are repressed in different kinds of cells.

Epigenetic Inheritance

It is clear that at least some epigenetic modifications are heritable, passed from parents to offspring in a phenomenon that is generally referred to as epigenetic inheritance, or passed down through multiple generations via trans-generational epigenetic inheritance. The mechanism by which epigenetic information is inherited is unclear; however, it is known that this information, because it is not captured in the DNA sequence, is not passed on by the same mechanism as that used for typical genetic information. Typical genetic information is encoded in the sequences of nucleotides that make up the DNA; this information is therefore passed from generation to generation as faithfully as the DNA replication process is accurate. Many epigenetic modifications, in fact, are spontaneously "erased" or "reset" when cells reproduce (whether by meiosis or mitosis), thereby precluding their inheritance.

Impact of Epigenetics on Biomedicine

Epigenetic changes not only influence the expression of genes in plants and animals but also enable the differentiation of pluripotent stem cells (cells having the potential to become any of many different kinds of cells). In other words, epigenetic changes allow cells that all share the same DNA and are ultimately derived from one fertilized egg to become specialized—for example, as liver cells, brain cells, or skin cells.

As the mechanisms of epigenetics have become better understood, researchers have recognized that the epigenome—chemical modification at the level of the genome—also influences a wide range of biomedical conditions. This new perception has opened the door to a deeper understanding of normal and abnormal biological processes and has offered the possibility of novel interventions that might prevent or ameliorate certain diseases.

Epigenetic contributions to disease fall into two classes. One class involves genes that are themselves regulated epigenetically, such as the imprinted (parent-specific) genes associated with Angelman syndrome or Prader-Willi syndrome. Clinical outcomes in cases of these syndromes depend on the degree to which an inherited normal or mutated gene is or is not expressed. The other class involves genes whose products participate in the epigenetic machinery and thereby regulate the expression of other genes. For example, the gene MECP2 (methyl CpG binding protein 2) encodes a protein that binds to specific methylated regions of DNA and contributes to the silencing of those sequences. Mutations that impair the MECP2 gene can lead to Rett syndrome.

Many tumours and cancers involve epigenetic changes attributable to environmental factors. These changes include a general decrease in methylation, which is thought to contribute to the increased expression of growth-promoting genes, punctuated by gene-specific increases in methylation that are thought to silence tumour-suppressor genes. Epigenetic signaling attributed to environmental factors has also been associated with some characteristics of aging by researchers that studied the apparently unequal aging rates in genetically identical twins.

One of the most promising areas of epigenetic investigation involves stem cells. Researchers have understood for some time that epigenetic mechanisms play a key role in defining the "potentiality" of stem cells. As those mechanisms become clearer, it may become possible to intervene and effectively alter the developmental state and even the tissue type of given cells. The implications of this work for future clinical regenerative intervention for conditions ranging from trauma to neurodegenerative disease are profound.

References

- Organelles: microscopemaster.com, Retrieved 14, March 2020
- Cell-division: biologydictionary.net, Retrieved 06, June 2020
- Epigenetics: britannica.com, Retrieved 27, February 2020

3

Membrane Biophysics

The study of principles related to biological membranes including their structural and mechanical characteristics is known as membrane biophysics. It also involves the study of protein lipid interactions and transmembrane helices, nature and magnitude of cell surface charge, etc. The topics elaborated in this chapter will help in gaining a better perspective about membrane biophysics.

A Membrane can be described as the boundary which define specific region from rest of the surrounding. It is not only gives a definite boundary but also control the composition of the enclosed space. In living systems each cell is covered by a 5 to 10 nm wide layer of phospholipids and proteins, which encloses whole cell and protects cell cytoplasm from extracellular environment and this biological membrane is 'cell-membrane' or commonly known as the plasma membrane. In addition to the plasma membrane, a eukaryotic cell contains various intracellular organelles like Golgi apparatus, mitochondria etc. and these organelles are surrounded by their intracellular membranes. These membrane coverings maintain the characteristic differences between the cell cytoplasm and contents of each respective organelle. Another important property of cell membrane is its semi-permeable nature, through which selective transport and retention of cell nutrients takes place and toxic or unwanted substances cannot pass into the cell.

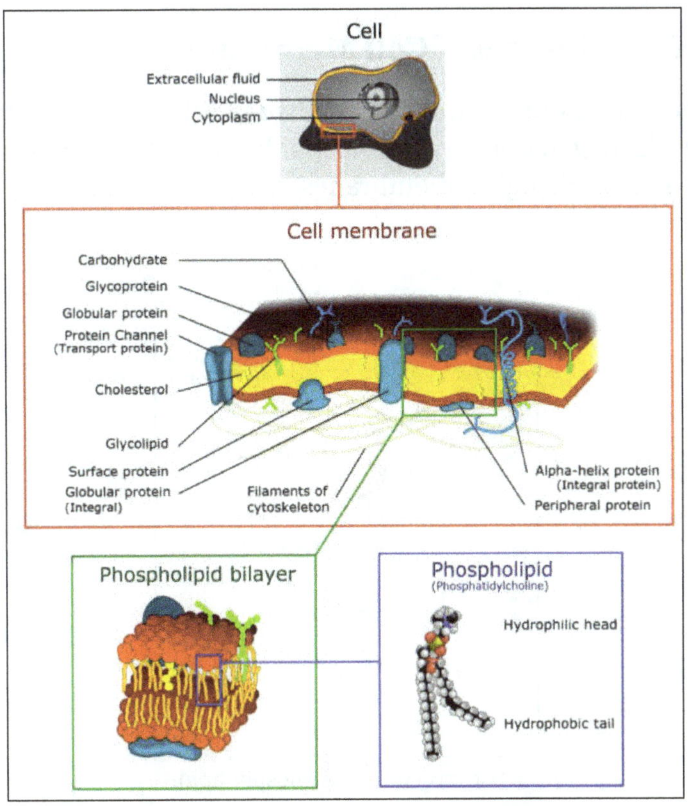

General and ultra-structure cell membrane.

Basic Components of Cell Membrane

Three bio-molecules namely lipid, protein and carbohydrates are the main components of a plasma membrane. Their compositions and relative proportions in membranes may vary in different types of cells. The variation in proportion of each component is provided for the special requirements of the cell for example in general plasma membranes have approximately equal proportions of lipids and proteins (45-50%). Whereas, mitochondrial inner membrane contains approximately 75% proteins which are required in electron transport chain. Lipids are present ubiquitously in different forms like phospholipids, cholesterol etc. in cell membranes and constitute approximately 50% of the mass of most cell membranes. Proteins may range from 20% to 70% of the total mass of a particular membrane. The percentage of carbohydrates in cell membrane is approximately 5-8% and these are generally associated with either to lipid (glycolipids) of various classes, or with proteins (as glycoproteins). A common feature of all bio-membranes is the presence of a bilayer of phospholipids; however, there is the presence of several unique proteins in certain cellular membranes for distinctive functions.

Table: Percentage distribution of lipids, proteins, and carbohydrates in different bio-membranes.

S.no.	Type membrane	% Lipid	% Protein	% Carbohydrate
1	Human erythrocyte plasma membrane	43	49	8
2	Amoeba plasma membrane	42	54	4
3	Mitochondrial inner membrane	24	79	0
4	Myelin	79	18	3

Lipids: The Main Constituents of Cell Membrane

The lipids components of membranes are further divided into glycerophospholipids, sphingolipids, glycolipids and sterols. In general, glycerophospholipids and sphingolipids constitute the largest proportions of lipids in all biological membranes.

Phospholipids

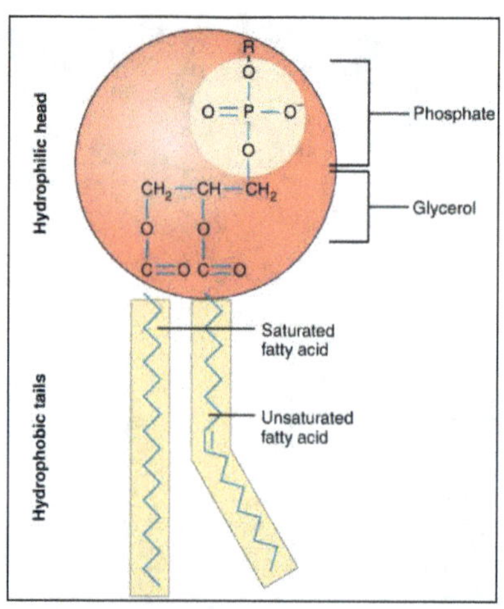

Basic structure of a phospholipid.

With few exceptions, the most abundant lipid components present in membranes are phospholipids. Phospholipids are basically composed of glycerol backbone at which two fatty acid chains and a phosphate group are attached. Fatty acids are attached to the first and second carbons, while the phosphate group is attached to the third carbon of the glycerol backbone. The fatty acyl side chains may be saturated or may contain one or more double bonds. The phosphate group and glycerol backbone makes the head portion of phospholipid as hydrophilic, whereas the fatty acid tails have a nature of hydrophobicity. Therefore, phospholipids are considered as amphipathic molecules with a polar head and a hydrophobic tails.

One more special characteristic of phospholipids make them a crucial molecule in membranes. The phosphate group of a phospholipid can be attached to variable functional groups(R group) and give rise to different types of phospholipids. For example Phosphatidic acid has the hydrogen as R group. The naming of phospholipid is based upon by naming the derivatives for the head-group alcohol with "phosphatidyl-" as a prefix. Phosphatidyl-choline, phosphatidyl-ethanolamine, and phosphatidyl-serine are other common membrane phospholipids having choline, ethanolamine and serine respectively. Table below is showing different types of phospholipids which are formed on the basis of modification of R group in phosphate.

Table: Phospholipids with different head groups.

S.n.	Name of R group	Formula of R	Phospholipid type
1	—	$-H$	Phosphatidic acid
2	Ethanolamine	$-CH_2-CH_2-N^+H_3$	Phosphytidyletahnolamine
3	Choline	$-CH_2-CH_2-N^+(CH_3)_3$	Phosphtidylcholine
4	Serine	$-CH_2-CH(COO^-)-\overset{+}{N}H_3$	Phosphtidylserine
5	Glycerol	$-CH_2-CH(OH)-CH_2-OH$	Phosphtidylglycerol
6	Myo-inositol 4,5, bisphosphate	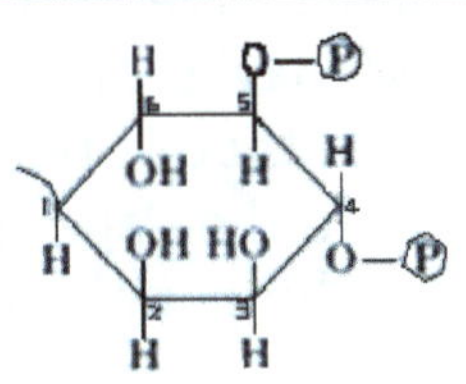	Phosphtidylinositol 4,5, bisphosphate
7	Phosphtidylglycerol	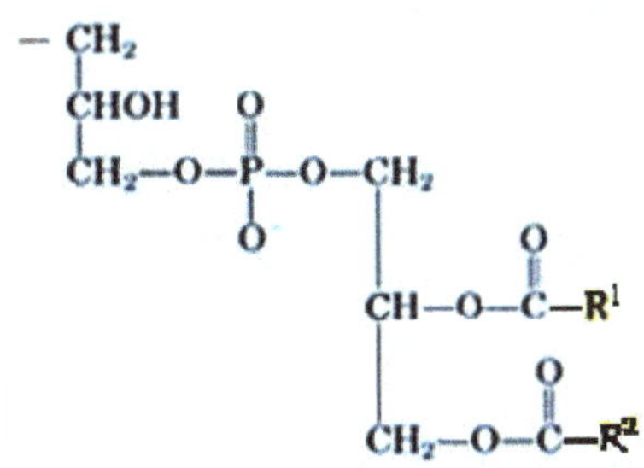	Cardiolipin

Phosphatidylinositol (PI) are generally present on the cytoplasmic leaflet of membranes and get phosphorylated into PI phosphate (PIP), PI bisphosphate (PIP_2), and PI trisphosphate (PIP_3). These phospholipids are important for regulating cell growth, their proliferation, and apoptosis. In general, majority of phospholipids are neutral in nature (e.g. phosphatidylcholine, phosphatidylethalamine and spingomyelin). However, several phospholipids like phophatidylionositol and phosphotidylserine shows acidic nature with negative charge.

When phospholipids are placed in an aqueous solution they will self-assemble into micelles or bilayers, structures that exclude water molecules from the hydrophobic tails while keeping the hydrophilic head in contact with the aqueous solution. The same phenomenon is observed in cell membrane where water content is present inside and outside of the cell and phospholipids are packed as bilayer confirmations. Moreover, the cylindrical shape of phospholipids, allows phospholipids to align side-by-side to form broad sheets.

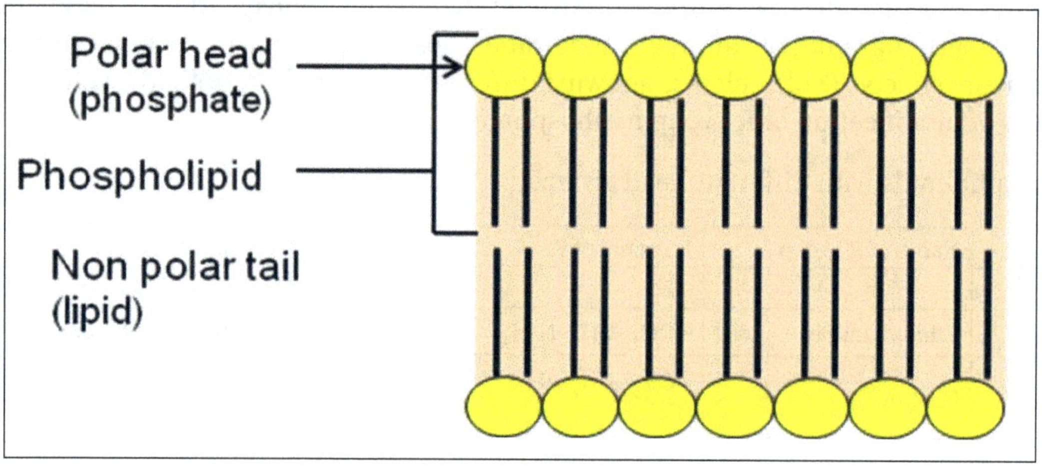

Orientation of phospholipids as bilayer in membrane.

Sphingolipids and Glycolipids

Another class of phospholipids found mainly in plasma membranes are sphingolipids which lack the basic glycerol backbone and have a sphingosine backbone fatty acid, and headgroup of a sphingolipid. In sphingolipids, the terminal hydroxyl group of sphingosine is esterified to phosphocholine, so its hydrophilic head is similar to that of phosphatidylcholine. Ceramide is the most basic sphingolipid which only a hydrogen in the Csp-3 position and an amide-linked fatty acid. Sphingomyelin is another most common sphingolipid, widely present in myelin sheaths of neurons.

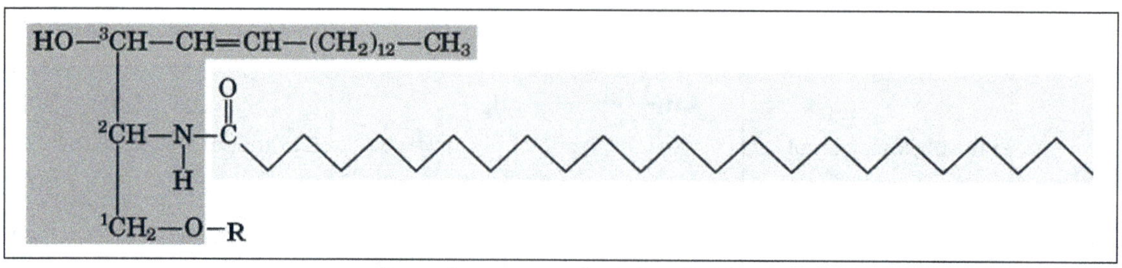

General structure of a Spingolipid.

Table: Spingolipids with different head groups.

S.n.	Name of R group	Formula of R	Spingolipids type
1	–	–H	Ceramide
2	Phosphocholine	PO_4^-–CH_2–CH_2–N^+ $(CH_3)_3$	Spingomyelin
3	Glucose	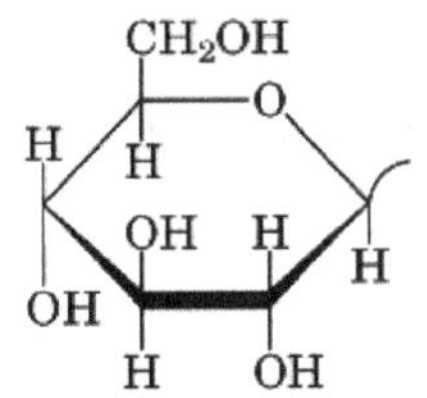	Glucosylcerebroside
4	Di/tri/tetrasaccharide	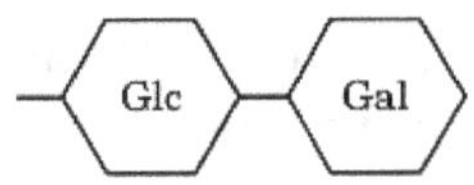	Lectosylceramide
5	Complex oligosaccharides	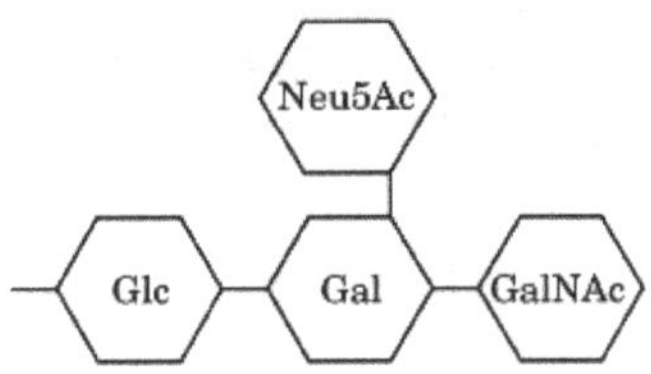	Ganglioside GM2

Glycolipids are the sub-group of sphingolipids and generally present approximately 2% of the total lipid content of plasma membranes. These sugar containing lipids have a spingosine derived back-bone and mono- or oligosaccharide bound to the Csn-3 position. For example, Glucosylcerebroside is a glycolipid, abundant in myelin sheaths and it consists of the ceramide and oleic acid linked to a single glucose residue. These are generally found at the outer leaflet of the plasma membrane and their carbohydrate moieties are exposed on cell surface.

Sterols

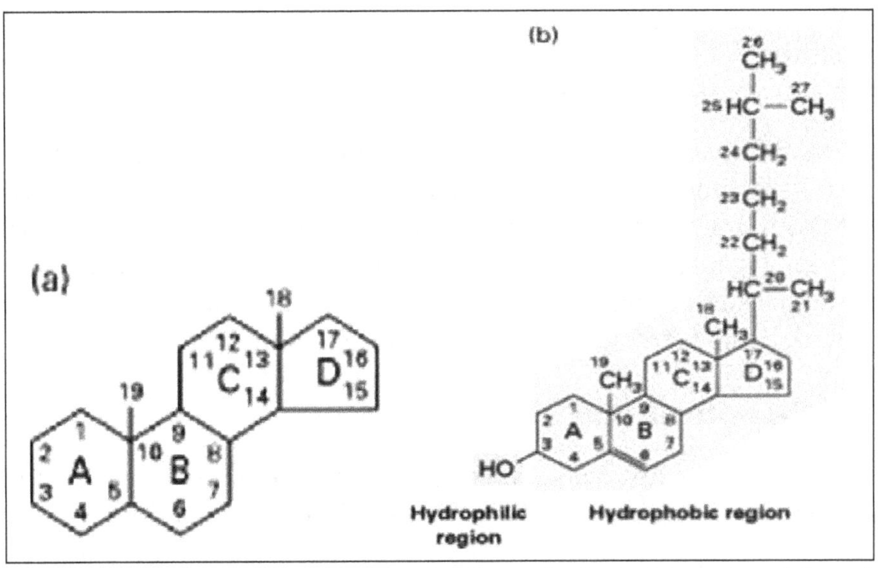

(a): The structure of sterols and (b) structure of cholesterol. The major portion of the molecule is hydrophobic (grey), the hydroxyl group is present at 3rd position.

Another important class of membrane lipids is the Sterols and their derivatives. These sterols are present in the cell membrane for important hormones and signaling molecules. In general, all steroids contain four hydrocarbon rings (A, B, C, and D), with the carbons numbered as shown figure. Cholesterol is the major steroidal constituent of plasma membranes. It accounts for near-about 20% and 30-50% of the of total lipid content in animal cell and plant cell, respectively. However, cholesterol is generally absent in prokaryotic membranes and mitochondrial membrane. Cholesterol has the entire hydrocarbon structure, but also has a hydroxyl substituent on one ring which makes this molecule as amphipathic. Cholesterol maintains the rigidity and stability of the membrane.

Membrane Proteins

Proteins also contributed the major proportion (50% of total mass of membrane) in biological membrane and these proteins participated in several biological activities like transport, enzymatic actions, communications etc. Structurally, membrane proteins may be present as single protein or as multimeric protein subunits. They may also get conjugated with carbohydrates and lipids and present in the form of glycoprotein and lipoprotein, respectively. Lipid moiety of a lipoprotein may have direct interaction with the lipid bilayer or through integral membrane proteins.

Classification of Membrane Proteins

Location based Classification

One the classification of membrane proteins is on the basis of their location in membrane. Membrane proteins are categorized as integral membrane proteins and peripheral membrane proteins:

- Integral membrane proteins: The proteins which are present in between the phospholipid bilayer or its major part penetrate or span the bilayer are termed as Integral membrane proteins or trans-membrane proteins. The membrane-spanning portions of transmembrane proteins are usually composed of hydrophobic amino acids which make αhelical secondary confirmations except porins (pore forming proteins in bacteria). In porins beta-sheets structures are predominant. In general, these proteins are tightly bound to the membrane through hydrophobic interactions and their terminal portions may exposed on one or both sides of the membrane.

- Peripheral membrane proteins: The proteins present outside the membrane are considered as peripheral membrane proteins. These proteins do not interact with the hydrophobic core of the bilayer and are loosely associated with the membrane through non-covalent interactions (generally ionic bonds with the polar head groups of the phospholipids). Therefore, it is easier to isolate peripheral membrane proteins and can be done by using the solutions of high ionic strength/salt concentration on membrane.

Function based Classification

Proteins present in membrane are responsible for the vital functions of cell and therefore these proteins may also be differentiated on the basis of their following functions:

- Transport or Channel Proteins: Gaseous molecules e.g. O_2 and CO_2 and small hydrophobic molecules can pass through lipid bilayer, but, water and other hydrophilic nutrients generally can't pass through lipid bilayer or at a very slow pace. The movement of the hydrophilic molecules is done by the transport proteins which are generally embedded in the membrane. These transport proteins shows specificity for selection of the molecules to be transported. In several occasions, these proteins transport nutrients against their concentration gradient and energy of ATP is utilized to catalyze passage. The ability to maintain concentration gradients and sometimes move materials against them is vital to cell health and maintenance. Thanks to membrane barriers and transport proteins, the cell can accumulate nutrients in higher concentrations than exist in the environment and, conversely, dispose of waste products.

- Carrier Proteins: These proteins are generally present inside membrane with some part surface exposed and binding site on protein surface "grabs" certain molecules and pulls them into the cell e.g. gated channels.
- Receptor Proteins: The cell membrane also contains some specific proteins which act as sensors (or molecular triggers) for external signals and allow the cell to produce response due to change in external environment of cell. These protein sensors are also termed as cell protein receptors. Their main function is to transfer information rather than transport of molecules or ions across the membrane e.g. release of hormones due to receptor proteins.
- Cell Recognition Proteins: Several proteins are present as glycoproteins, where attached carbohydrates are localized on the outer exo-plasmic membrane face of eukaryotic cells. These proteins act as ID tags for identification of specific cells by the immune system of body.
- Enzymatic Proteins: Several proteins present on the membrane surface shows enzymatic activities and carry out metabolic reactions.

Structure of the Cell Membrane

A cell's plasma membrane defines the cell, outlines its borders, and determines the nature of its interaction with its environment. Cells exclude some substances, take in others, and excrete still others, all in controlled quantities. The plasma membrane must be very flexible to allow certain cells, such as red and white blood cells, to change shape as they pass through narrow capillaries. These are the more obvious plasma membrane functions. In addition, the plasma membrane's surface carries markers that allow cells to recognize one another, which is vital for tissue and organ formation during early development, and which later plays a role in the immune response's "self" versus "non-self" distinction.

Among the most sophisticated plasma membrane functions is the ability for complex, integral proteins, receptors to transmit signals. These proteins act both as extracellular input receivers and as intracellular processing activators. These membrane receptors provide extracellular attachment sites for effectors like hormones and growth factors, and they activate intracellular response cascades when their effectors are bound. Occasionally, viruses hijack receptors (HIV, human immunodeficiency virus, is one example) that use them to gain entry into cells, and at times, the genes encoding receptors become mutated, causing the signal transduction process to malfunction with disastrous consequences.

Fluid Mosaic Model

The fluid mosaic model, has evolved somewhat over time, but it still best accounts for plasma membrane structure and function as we now understand them. The fluid mosaic model describes the plasma membrane structure as a mosaic of components—including phospholipids, cholesterol, proteins, and carbohydrates—that gives the membrane a fluid character. Plasma membranes range from 5 to 10 nm in thickness. For comparison, human red blood cells, visible via light microscopy, are approximately 8 μm wide, or approximately 1,000 times wider than a plasma membrane. The membrane does look a bit like a sandwich.

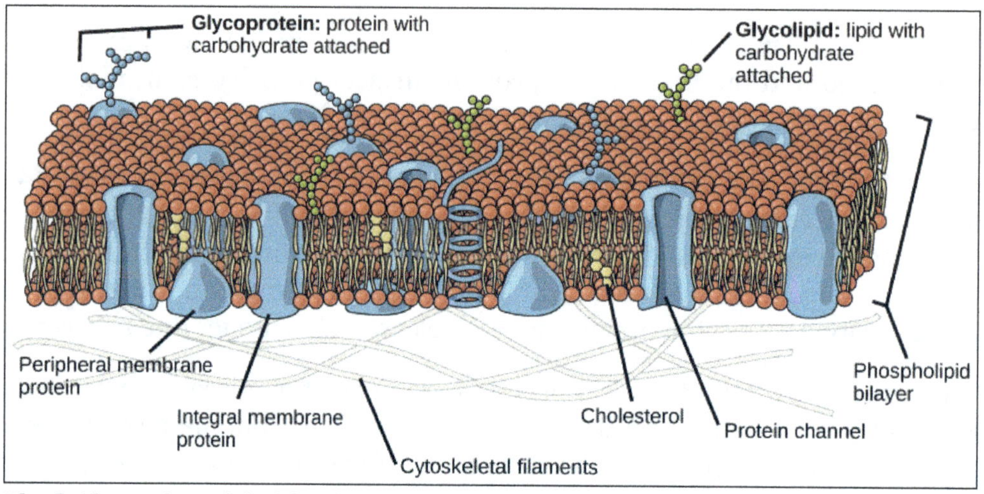

The fluid mosaic model of the plasma membrane structure describes the plasma membrane as a fluid combination of phospholipids, cholesterol, proteins, and carbohydrates.

A plasma membrane's principal components are lipids (phospholipids and cholesterol), proteins, and carbohydrates attached to some of the lipids and proteins. A phospholipid is a molecule consisting of glycerol, two fatty acids, and a phosphate-linked head group. Cholesterol, another lipid comprised of four fused carbon rings, is situated alongside the phospholipids in the membrane's core. The protein, lipid, and carbohydrate proportions in the plasma membrane vary with cell type, but for a typical human cell, protein accounts for about 50 percent of the composition by mass, lipids (of all types) account for about 40 percent, and carbohydrates comprise the remaining 10 percent. However, protein and lipid concentration varies with different cell membranes. For example, myelin, an outgrowth of specialized cells' membrane that insulates the peripheral nerves' axons, contains only 18 percent protein and 76 percent lipid. The mitochondrial inner membrane contains 76 percent protein and only 24 percent lipid. The plasma membrane of human red blood cells is 30 percent lipid. Carbohydrates are present only on the plasma membrane's exterior surface and are attached to proteins, forming glycoproteins, or attached to lipids, forming glycolipids.

Protein Lipid Interactions and Transmembrane Helices

The interplay of membrane proteins and lipids facilitates basic processes of respiration, photosynthesis, protein and solute transport, signal transduction and motility. This broad range of functions in the cellular metabolism is covered by a large number of diverse membrane-embedded proteins, which account for the coding capacity of 20–30% of the genes in a typical organism.

Three types of binding modes have been defined for lipid interactions with membrane proteins. (i) A shell of 'annular lipids' bound to the protein surface resembles the bilayer structure. (ii) 'Non-annular surface lipids' are immersed in cavities and clefts of the protein surface. They are frequently observed for multi-subunit complexes and multimeric assemblies and are typically present at contact sites between adjacent subunits or monomers. (iii) 'Integral protein lipids' reside within a membrane protein or a membrane protein complex. They often occur in unusual positions, e.g. with the head group below the membrane plane and/or the acyl chains non-perpendicular to it.

The lipid bilayer provides in general the matrix for the membrane proteins, but a tight protein–lipid interaction is required to maintain the diffusion barrier and to keep it electrochemically sealed. The mobile lipid molecules can adhere to the membrane protein surface and flexibly adjust to conformational changes and structural rearrangements. Protein–lipid interactions are specific; numerous biochemical and biophysical studies have demonstrated that defined lipid species confer structural stability, control insertion and folding processes, are involved in assembly or oligomerization of multi-subunit complexes or super complexes or directly affect the function of membrane proteins.

The increasing numbers of membrane protein structures, which in part contain tightly bound lipids, pave the way to analyze the role of specific lipid binding sites for the structure and function of membrane proteins. Protein-lipid interaction is the influence of membrane proteins on the lipid physical state or vice versa. The questions which are relevant to understanding of the structure and function of the membrane are: 1) Do intrinsic membrane proteins bind tightly to lipids, and what is the nature of the layer of lipids adjacent to the protein? 2) Do membrane proteins have long-range effects on the order or dynamics of membrane lipids? 3) How do the lipids influence the structure and/or function of membrane proteins? 4) How do peripheral membrane proteins which bind to the layer surface interact with lipids and influence their behavior?

Transient lipid protein interaction is the a particular case of lipid-protein interaction which is represented by complement, perforins, several viruses and protein toxin that are water soluble and yet under certain conditions become able to penetrate into the hydrophobic core of the lipid bilayer. It has also been shown that the acidic pH reached in the endosomes, where these proteins are collected via receptor-mediated endocytosis, induced a structural change with exposure of hydrophobic surface. Thus the protein penetrates into the endosomal lipid bilayer and in case of viruses, induces the fusion of the viral and endosomal membrane.

Integral proteins interact both with the polar head group of phospholipids and with fatty acid chain, while peripheral proteins are thought to be involves in a interaction with the polar head group only and are not expected to perturb significantly the hydrophobic region of the lipid bilayer. Despite their large non-homogeneity, due to presence of variable number of double bonds in different positions, hydrocarbon chains are expected, for the very nature of the forces involved, not to show any particular specific interaction with any membrane protein. They simply provide a plastic, dynamic and water repelling structure and seal the membrane around integral proteins by hydrophobic interactions with the central surface of the hydrophobic sector formed by non-polar amino acid lateral residues.

There are various forces are involved in the lipid-protein interaction. The non-covalent forces which are mainly involved in the stability of membrane or lipid protein interaction are hydrophobic force and van der Waals forces. The hydrophobic interactions are non-covalent interactions that minimize contact of hydrophobic or non-polar molecule with the aqueous hydrophilic or amphipathic molecules i.e. it provide tendency for nonpolar molecules to aggregate in solution. Hydrophobic amino acids such as Ile, Val, Leu, Phe, Met and Cys play a significant role in the lipid interaction. Similarly, van der Waal's force can be formed by every atom involved in the protein lipid interaction. Van der Waal's force is a non-directional weak electrostatic attractive or repulsive force formed between electrically neutral species. Therefore, every atom of the protein and lipid can form the van der Waals force.

Mode of Binding of Lipids to Membrane Proteins and Specific Lipid-Binding Sites

A large research effort involves approaches to know whether proteins have binding sites which are specific for particular lipids and whether the protein–lipid complexes can be considered to be long-lived, on the order of the time required for the turnover a typical enzyme that is 10^{-3} sec. This is now known through the use of 2H-NMR, ESR, and fluorescent methods.

The lipid is not as well oriented as is the bulk lipid, and it can occupy a broader distribution of orientations; it is disordered. The rough surface presented by a membrane protein to the surrounding lipid bilayer will result in poor packing unless the lipid fatty acyl chains distort to match the surface of the protein, and poor packing must be avoided to maintain the membrane as an effective permeability barrier. The presence of a rigid protein surface will reduce the extent of the motional fluctuations of the chains and the chains will have to tilt and become conformational disordered to maximize contact with the surface of the protein. The hydrophobic surface of a membrane protein will therefore be covered by a shell of disordered lipids, referred to as boundary or annular lipids, the latter term referring to the fact that the lipid molecules form a ring or annulus around the protein.

Specific Lipid-Binding Sites

The residues surrounding the lipid fatty acyl chains are hydrophobic, Trp residues being common near the tops of the chains, with Ile, Leu, Phe, and Val commonly surrounding the rest of the chain. Good van der Waals contact between the chains and the protein is often ensured by binding of at least the top parts of the chains in distinct groves on the surface of the protein, but this may not be true for all of the bound lipid molecules; it may be that only those lipid molecules whose chains are bound in grooves are sufficiently immobilized to be resolved in the X-ray structures. In many of the structures, lipid headgroups are disordered, suggesting that the headgroups interact less tightly with the protein than the lipid fatty acyl chains; examples include bacteriorhodopsin and the potassium channel KcsA. Hydrophobic thicknesses for membrane proteins whose crystal structures do not include any resolved lipid molecules can be estimated from the locations of aromatic and charged residues. The crystal structure of the light harvesting complex from Rhodopseudomonas acidophila shows the importance of charged residues in locating the membrane surface.

Assays or Methods to Study Protein-Lipid Interactions

To understand the molecular mechanism by which a protein interacts with biological membranes, one needs experimental knowledge of the affinity and specificity of the membrane-binding process and the protein's topology/conformation in the lipid bilayer, as well as possible modulation of physicochemical properties of the membrane by proteins. Thus many methods, including spectroscopy, microscopy, and cosedimentation and co-flotation assays, have been developed to aid understanding of the principles of interactions between lipids and proteins. Here we will discuss few methods such as Flotation assay, ITC (isothermal titration calorimetry, SPR (surface plasmon resonance), MST (microscale thermophoresis), Native protein mass spectrometry and recently developed NMR and Fluorescence methods.

Flotation assay is a method in which liposomes and proteins are mixed at the bottom of a sucrose gradient and ultracentrifuge. If proteins and lipids interact, the complex floats in the upper fractions of the centrifugation tube. ITC (isothermal titration calorimetry) is a biophysical method to obtain thermodynamic parameters of the interaction between proteins and liposomes, which allows their molecular affinities to be calculated. In SPR (surface plasmon resonance) approach, a lipid membrane is formed or liposomes are bound on top of a plasmon resonance sensor chip, and SPR is used to detect the binding and dissociation of proteins. In MST (microscale thermophoresis) method, a temperature field is used to perform biomolecular interaction studies, which can be applied to protein–lipid interaction studies. Native protein mass spectrometry is used to analysis of intact protein complexes by mass spectrometry.

Membrane Proteins and Classification of Transmembrane Helix

Increasingly, drug developers are looking to large molecules & proteins as a therapeutic option. Plan of a protein drug product can be a significant test, however without a decent comprehension of the nature of protein structure and the conformational characteristics of the particular protein being formulated, the outcome can be harmful. Proteins are macromolecules and have four distinct levels of structure- primary, secondary, tertiary and quaternary. The primary structure of a protein is the linear sequence of its amino acid structural units. The primary structure of a protein is accounted for beginning from the amino-terminal (N) end to the carboxyl-terminal (C) end. Protein secondary structure is the general three-dimensional type of local segments of proteins. Secondary structure can be formally characterized by the example of hydrogen bonds of the protein, (for example, alpha helices and beta sheets) that are seen in a atomic-resolution structure.

The tertiary structure will have a single polypeptide chain "backbone" with one or more protein secondary structures, the protein domain. Quaternary structure is the number and arrangement of multiple folded protein subunits in a multi-subunit complex. It may have associations from simple dimers to large homo-oligomers and complexes with characterized or variable quantities of subunits. The secondary and tertiary structure of proteins are embedded in the membrane which have some trans-membrane region. Trans-membrane receptor protein consists of:

- The Extracellular Domain of Transmembrane Receptors.
- The Transmembrane Domain.
- The Intracellular Domain of Membrane Receptors.

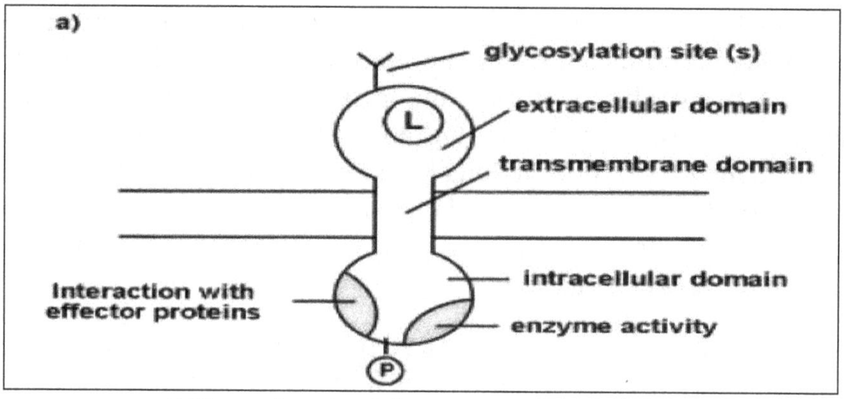

Different domain of the trans-membrane helix.

There are following classes of trans-membrane alpha helixes found in the membrane proteins:

- Single transmembrane helix,
- Multi-transmembrane helices,
- Beta-barrel or channel,
- Anchored by one (not-transmembrane) helix or a covalently attached fatty acid.

Transmembrane segment ordinarily means a single transmembrane alpha helix of a transmembrane protein, also called an integral protein. More extensively, a transmembrane segment is any three-dimensional protein structure which is thermodynamically steady in a membrane. This might be a single alpha helix, a transmembrane beta barrel, a beta-helix of gramicidin A, or any other structure. Transmembrane helices in integral membrane proteins are made out of extend of 15–30 dominatingly hydrophobic residues isolated by polar connecting loops. Learning of the presence and exact location of the transmembrane helices is vital for useful explanation and to coordinate functional analysis. Various algorithms designed to locate putative transmembrane helices in the primary amino acid sequence have been produced, and current strategies can distinguish around 90–95% of all genuine transmembrane segments with an over production rate of just a couple percent. Transmembrane helices are substantially easier to predict than helices in globular domains. Predicting 95% of the transmembrane helices in the 'correct' location is not unusual.

Methods for prediction of transmembrane helices are usually based on (1) hydrophobicity analyses (2) the preponderance of positively charged residues on the cytoplasmic side of the transmembrane segment (interior), established as the "positive inside rule" or (3) statistical procedures that perform significantly better when combined with multiple alignments. But may be detailed aspects of 3D structure be predicted from sequence for HTM proteins?

Prediction of Transmembrane Helices Proteins

Methods for prediction of transmembrane helices are usually based on:

- Hydrophobicity analyses.
- The preponderance of positively charged residues on the cytoplasmic side of the transmembrane segment (interior), established as the "positive inside rule".

- Statistical procedures that perform significantly better when combined with multiple alignments.

Cytoplasmic and extracellular regions have different amino acid compositions. This difference allows for a successful prediction of not only the location of helices but, as well, of their orientation with respect to the cell (pointing inside or outside the cell). Going further, Taylor and colleagues enumerate all possible models for packing seven helix transmembrane proteins and select the "better models". The selection criterion for "better models" is the crucial point of the method. In a report presented that the native conformation is found in "most cases" tested. However, the N- and C-terminal ends of the transmembrane helices have to be predicted very accurately for a successful automatic prediction of 3D structure from sequence.

Prediction accuracy has recently been improved significantly. A system of neural networks using single sequences as input appears to be slightly inferior to these methods. However, using information from multiple sequence alignments as input, neural networks have been shown to yield the most accurate prediction of secondary structure for globular proteins. We can send a similar system of neural networks to predict transmembrane helices based on evolutionary information. The goal was to predict the location of transmembrane helices (defined as helix caps given in SWISS-PROT more accurately than alternative methods.

A method to predict the complete topology of transmembrane proteins, TopPred, applies two empirical hydrophobicity cutoffs to the output of a sliding trapezoid window in order to compile a list of certain and putative transmembrane helices. The combination of putative helices that produces the strongest enrichment of positively charged residues in loops on the cytoplasmic side is selected as the best prediction. Loops that are longer than 70 residues are ignored. Drawback of this method and other method is that depend on fixed hydrophobicity thresholds for considering a segment as a transmembrane helix is that some helices may be missed that fall just under the threshold. This is not unusual in proteins with many membrane-spanning helices that form a bundle in which non-hydrophobic residues may make contacts between helices.

Transmembrane helices are visible in structures of membrane proteins determined by X-ray diffraction. They might likewise be predicted on the basis of hydrophobicity scales. Since the interior of the bilayer and the interior of most proteins of known structure are hydrophobic, it is attempted to be a required of the amino acids that span a membrane that they hydrophobic too. Membrane pumps and ion channels likewise contain numerous charged and polar residues within the generally non-polar transmembrane segments. Using hydrophobicity analysis to predict transmembrane helices empowers an prediction thus of the "transmembrane topology" of a protein; i.e. prediction of what parts of it protrude into the cell, what parts protrude out, and how often the protein chain crosses the membrane. Such prediction techniques are generally connected with a limited success.

The best results have been acquired when multiple aligned sequences can be analyzed; however, much of the time there are no homologues in singlesequence prediction performance are in this way vital. Dense alignment surface (DAS) method was introduced in an attempt with enhance sequence alignment in the G-protein coupled receptor family of trans-membrane proteins. This method is generalized to predict transmembrane segments in any integral membrane protein without the need for multiple sequence information and also found that it performed on par with the best multiple-alignment based schemes when tested on a set of prokaryotic inner membrane proteins with known topologies.

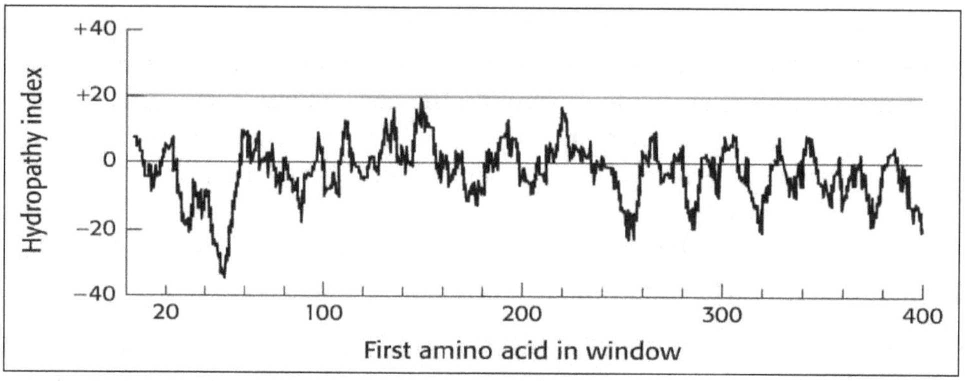

Prediction of transmembrane regions of membrane proteins by hydropathy plots.

Drawback of Hydropathy Plot

Although hydrophobic plot can give information about hydrophobic property of the each amino acid in the membrane protein which will help to understand the location of each amino acid in the membrane protein for example in channel forming protein a hydrophobic amino acid will always present towards the membrane side and a hydrophilic amino acid is always present in the core but it also have some drawback. This plot also has following failures such as:

- False positive: Hydrophobic parts of globular proteins.
- Beta-barrel structures.
- Helices including a highly hydrophilic surface.
- Assembly of transmembrane helices cannot be predicted.

Nature and Magnitude of Cell Surface Charge

In terms of physics, a 'surface charge' may be defined as the electrical potential difference generated between the inner and outer surface of the dispersed phase in a colloid. Different processes like ion adsorption, protonation/deprotonation and an external electric field can make a charged surface. Surface charge appears on the surface by placing it in an electrolytes containing fluid (it contains cationic and anionic species). These ions interact with the object surface which may lead to the adsorption of some of them onto the surface. Interaction and adsorption of ions make the surface ionized and develops a net surface charge on the surface.

In biological world, cell surface charges are most important in respect to understanding the nature of a bio-membrane and function associated with it. The first and foremost aspect of charge distribution comes from membrane architect. Membrane primarily consists of two layered structures of phospholipids. Each phospholipid has two ends in which one phosphate group is attached with the other a hydrocarbon chain. Both of these two moieties have very different properties.

The phosphate end is hydrophilic while the lipid chain has hydrophobic nature. Therefore, the arrangement of membrane phospholipids in water occur in such a way that they achieve a lower energy. Each molecule attempts to get into the lowest-energy state possible and due to their collective effort, a double-layered sheet with the lipid chains in the center and phosphate ends towards water side is formed.

Electric charges on membrane surfaces are important factors in restoration of electromagnetic field. If ions bound weakly on membrane surface, they are free to move hence, less effected by electromagnetic forces. However, ion bound robustly on membrane surface, electromagnetic field may cause alteration in their structure. It is a vital parameter in understanding membrane binding mechanisms and liposome associated drug delivery etc.

Surface charges on cell membrane are also associated with passive transport of charged molecules as it is affected by electrostatic forces and flux rates are dependent on intra-membrane field forces.

Conductance and Capacitance of Biological Membranes

Phospholipids are used to create the artificial membranes and different electrical properties are realized through these structures. Pure phospholipid bilayers are observed as insulator as there are no free ions in these bilayer membranes and hence no carriers are present to transport charges. Their specific conductance per unit area is approximately:

$$g_{pure} = 10^{-13}\Omega^{-1}m^{-2}$$

However, the conductance of biological membranes is much higher, typically by several orders of magnitude even at rest. The reason behind this is the presence of different kinds of ion channels, pumps and other pores which span the membrane and allow additional currents to flow.

The cytosol (inside) and the extra cellular medium (outside) of the cell are having a type of electrolyte mediums. These two regions constitute a good conductive environment because there are free ions that can transport electrical charges. The polar head groups of membrane phospholipids also participate as electrically conducting material with the surrounding aqueous medium. However, the hydrophobic interior part acts as an insulator. Therefore, the plasma membrane acts like a capacitor in which the two conductors (the inside and the outside of the cell) are separated by an insulator (the membrane). This also results in having different amounts of electrical charges inside and outside the cell.

If we can separate a charge Q by applying an electrical potential V across the membrane, the membrane has by definition a capacitance,

$$C = Q/V$$

As the membrane is only two molecules thick (approximately 6×10−9m total thickness), a small voltage is required to separate the charges and therefore the membrane capacitance is quite high; per unit area, it is,

$$c=C/S\approx 10^{-2}Fm^{-2}$$

where F is "Farad"; the unit of capacitance.

The specific capacitance of biological membranes is very close to the dielectric constant of lipids and the thickness of the bilayer. Unlike the conductance, the capacitance is very little influenced by all the complexities of biology.

Membrane Potential

Lipid bilayer of membrane serves as both an insulator and a diffusion barrier to the movement of ions. However, ion transporter proteins embedded in membrane actively transport ions across the membrane and establish concentration gradients across the membrane, and ion channels allow movement of ions across the membrane down those concentration gradients. Ion pumps and ion channels are electrically equivalent to a set of batteries and resistors inserted in the membrane, and therefore create a voltage between the two sides of the membrane. This voltage is generally termed as membrane potential (also known as transmembrane potential or membrane voltage) and is defined as the difference in electric potential between the interior and the exterior of a biological cell. Membrane potential (Vm) is expressed in millivolts. Membrane potential is responsible for producing an opposing force for ion movements. In general, all eukaryotic cells maintain a negative voltage in the cell interior as compared to the cell exterior which ranges from −40 mV to −80 mV.

The membrane potential serves two crucial functions in cell maintenance and communications.

- Due to membrane potential, cell functions as a battery and it provide power to operate a variety of transporting proteins embedded in the membrane.
- Neurons and muscle cells utilize the membrane potential for transmitting signals between different parts of a cell.

Opening or closing of ion channels in the membrane produce a local change in the membrane potential and a signal is generated. The change in the electric field can be quickly affected by either adjacent or more distant ion channels in the membrane. These affected ion channels behave according to the potential change and reproduces the signal.

A baseline stable membrane potential known as 'resting potential' of magnitude of −70 to −80 millivolts remains in the interior of a cell (e.g. neurons). The opening and closing of ion channels can induce a departure from the resting potential and if the interior voltage becomes less negative (from −70 mV to −60 mV), it is termed a depolarization and if the interior voltage becomes more negative (−70 mV to −80 mV), it is termed as hyperpolarization. An action potential is generated by the activation of several voltage-gated ion channels in excitable cells due to a large depolarization. During action potential, the membrane potential changes rapidly for a short span of time (1-100 milliseconds) and reverse its polarity. In neurons, chemical gated and voltage-gated ion channels influence the membrane potential. Since, voltage-gated ion channels and membrane potential are interlinked to each other for their control; a feedback loops that allow oscillations and regenerative events like action potentials.

Electrochemical Gradient/Potential Across the Membrane

The transportation of a solute across a membrane is dependent upon its chemical gradient (i.e. the difference in solute concentration). Therefore, the direction of spontaneous movement of a charged solute across a membrane is dependent on the electrical gradient as well as on the chemical gradient. Combination of both of these factors is referred as the electrochemical gradient/ potential. Electrochemical potential can be expressed as,

$$\mu_i' = \mu_i + z_i F \varphi$$

where:

- μ_i' is the electrochemical potential of species i, in J/mol.
- μ_i is the chemical potential of the species i, in J/mol.
- z_i is the valency (charge) of the ion i, a dimensionless integer.
- F is the Faraday constant, in C/mol.
- φ is the local electrostatic potential, in V.

In the special case of an uncharged atom, $z_i = 0$ and so $\mu_i = \mu_i$.

Electrochemical potential is important in molecular diffusion across membranes. It represents one of the many interchangeable forms of potential energy through which energy may be conserved.

A Model of the Phospholipid Membranes

The dielectric permittivities of the membrane and water are correspondingly: $\varepsilon_m \approx 2$ and $\varepsilon_w \approx 80$. It would seem that in order to evaluate the force acting on the ion of charge q near the membrane surface the formula can be applied:

$$F_q = \frac{q^2}{4h^2} \frac{\varepsilon_w - \varepsilon_m}{\varepsilon_0 (\varepsilon_w + \varepsilon_m)}$$

Here ε_0 is the permittivity of free space, and h is the distance to the interface between dielectrics. However, the paradoxical result follows from the formula above that the ion cannot get from the dielectric with the higher dielectric permittivity into the dielectric with the lower dielectric permittivity, because the force acting on it increases inversely proportional to the square of the distance to the boundary between dielectrics. This statement contradicts the experimental facts and theoretical models of Hodgkin and Huxley's kind, based on the phospholipid membrane permeability to ions. Usually, in theories of the interaction of ions with the surface of dielectrics, the minimum value of h is selected as the size of the ion, or the Debye radius, if the insulator's surface is saturated with ions. However, this does not resolve the problem of phospholipid membranes "impenetrability". In fact, the problem lies in that the formula above is derived in the approximation of continuous medium with the dipoles of the infinitely small size, without taking into account the real structure of the surface layer of the membrane.

It is known that the phospholipid molecules of cell membrane are forming a mosaic (matrix) structure in which dipole heads are directed towards the liquid (positively charged head faces outward membrane). The average surface area per molecule of the lipid is $\approx 0.5\text{nm}^2$, the length of the polar head is ~ 0.5–1nm, the radius of the head is ~ 0.2–0.3 nm, and the distance between the hydrophilic heads of the membrane is in the range of 5-7 nm. The dipole moment of the phospholipid head is

18.5-25 D. (1 D =$3.34 \cdot 10^{-30}$ Cm), i.e. more than 10 times greater than the dipole moment of water molecules. On the basis of geometrical dimensions of the cell membrane and the size of water molecules, it can be concluded that the free space between the head does not exceed the size of a water molecule (~ 0.2nm). That is, the membrane, interacting with the ions of surrounding liquid, cannot be considered as a dielectric medium with an infinitesimal dipoles size.

These facts allow the consideration of the following simplified model of the ion interaction with the membrane:

- The membrane represents a matrix with a mesh size a × a. In the nodes of cells the dipoles are located; the dipole charge is q; the distance between the charges (the dipole length), d; the distance between the dipoles along the axis z, l.
- An ion is a classical particle and cannot approach a dipole at a distance less than the sum of the radii of the head and the size of the ion. It is important that the ion can approach the membrane dipole heads close enough that would be "captured" by the potential well. This is a standard assumption in the theory of the interaction of ions with the surface of the dielectric. Since the dipole moment of water is 10 times less than the dipole m ment of the phospholipid head molecule and near the head cannot be more than one-two water molecules, the interaction between the ions located near the surface of the membrane and the water molecules can be neglected, as compared with the interaction of the ions with the dipoles of the membrane.
- We want to emphasize one more time that ionic permeability membrane is a separate problem and not consider in this work, for us it is important to know how strong ions attached to surface of membrane.

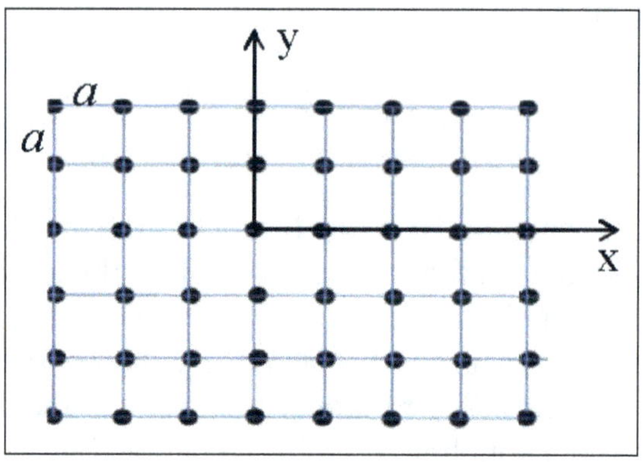

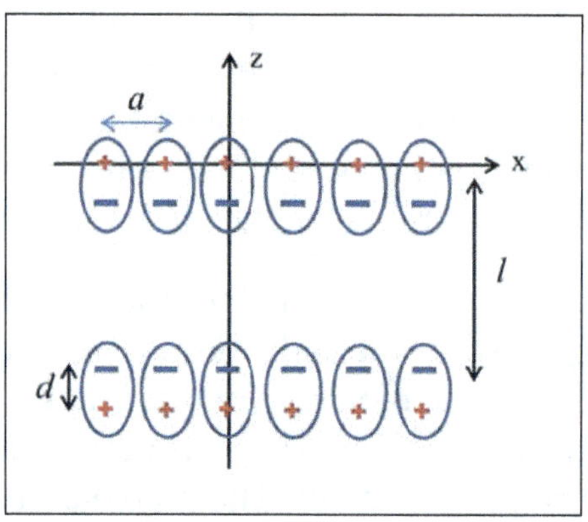

Figure shows a simplified mosaic model of a membrane: The membrane is a matrix with dipoles in the nodes. *a* is the mesh size of the matrix, *d* is the distance between the charges of the dipole head, *l* is the distance between the dipoles along the z axis.

The main ions in the physiological solution are: Na^+, K^+, Ca^{2+}, Mg^{2+} and Cl^-. The total concentration of these: $n_i \approx 1.8 \cdot 10^{26}$ m^{-3} (300mmL/Mole), but for simplicity we consider only the case when the bilayer phospholipid membranes are immersed into the water solution of NaCl.

The Potential Near the Surface of the Membrane

The expression for the membrane potential in a cylindrical coordinate system (r,θ,z) at a point above the surface of the membrane:

$$U(r,\theta,z)=\sum_{i=-n}^{n}\sum_{j=-n}^{n}U_{i,j}(r,\theta,z)=$$

$$\sum_{i=-n}^{n}\sum_{j=-n}^{n}\frac{q}{4\pi\varepsilon_0}\left(\frac{1}{\sqrt{r^2+r_{i,j}^2-2r\cdot r_{i,j}\cos(\theta-\theta_{i,j})+z^2}}\right.$$

$$-\frac{1}{\sqrt{r^2+r_{i,j}^2-2r\cdot r_{i,j}\cos(\theta-\theta_{i,j})+(z+d)^2}}$$

$$-\frac{1}{\sqrt{r^2+r_{i,j}^2-2r\cdot r_{i,j}\cos(\theta-\theta_{i,j})+(z+l)^2}}$$

$$\left.+\frac{1}{\sqrt{r^2+r_{i,j}^2-2r\cdot r_{i,j}\cos(\theta-\theta_{i,j})+(z+l+d)^2}}\right)$$

Here, the radial and angular coordinates $r_{i,j}$,,$\theta_{i,j}$ correspond to the position of the dipole in the node with the numbers (i, j), q is the dipole charge, and o ε is permittivity of free space. The value of n in equation above is chosen such that the potential near the membrane does not depend on the size of the matrix n. The count goes from the node (dipole head):

$r = (x^2 + y^2)^{1/2}$, $\theta = \arctan(y/x)$.

We expand the potential U in a Fourier series by θ. Taking into account the rotational symmetry on π/2 rotations for the accepted square matrix of the dipole heads, we obtain:

$$U(r,\theta,z)=U_o(r,z)+\sum_{j=1}^{j}U_j(r,z)\cos(4j\theta)+\sum_{j=1}^{j}V_p(r,z)\sin(4j\theta)$$

$$U_o(r,z)=-\int^{/2}U(r,\theta,z)d\theta \qquad V_o(r,z)=0$$

$$U_{j\neq o}(r,z)=\frac{4}{\pi}\int_o^{\pi/2}U(r,\theta,z)\cos(4j\theta)d\theta \quad V_{j\neq o}(r,z)=\frac{4}{\pi}\int_o^{\pi/2}U(r,\theta,z)\sin(4j\theta)d\theta$$

Spatial lattice of dipoles in bilayer lipid membrane. Potential is calculated at the point P. Ellipsoids at lattice sites correspond to the dipole heads of phospholipid molecules.

In the framework of the model presented above, estimates show that the influence of the dipoles located on the lower side of membrane on the potential distribution U at $z > 0$ can be neglected. The same is true to the effect of dipoles in the upper side of the membrane on the potential at $z<-(l+d)$. As far as the dependence of U on, in the region $r \leq a/2 = 0.35$nm the contribution of above the zero harmonics does not exceed 7%, so we will henceforth neglect them.

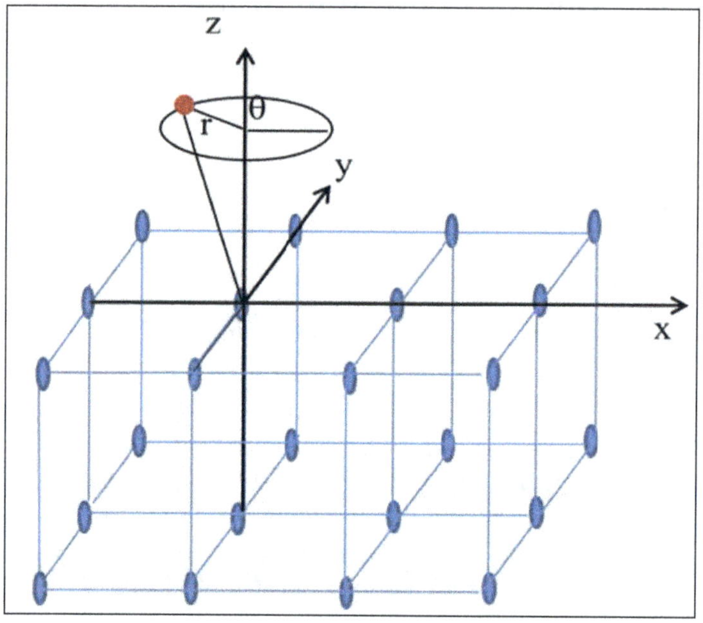

For example, let consider the case, when bilayer phospholipid membrane is immersed in the NaCl solution, at the following set of parameters: a = 0.7 nm, d = 0.5nm, l = 8 nm, and $q = 1.6 \cdot 10^{-19}$ C.

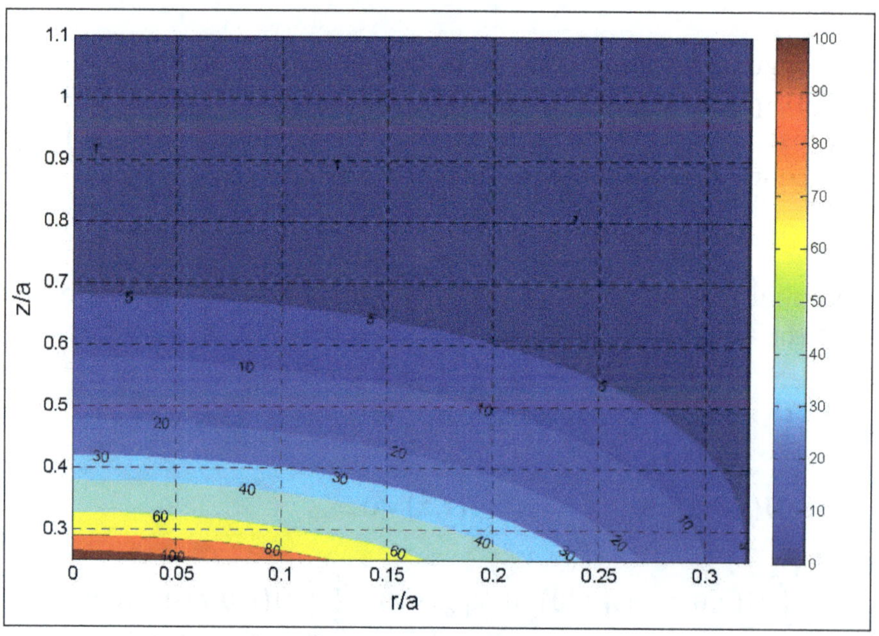

Contour plot U/kBT corresponding to a half of the cell.

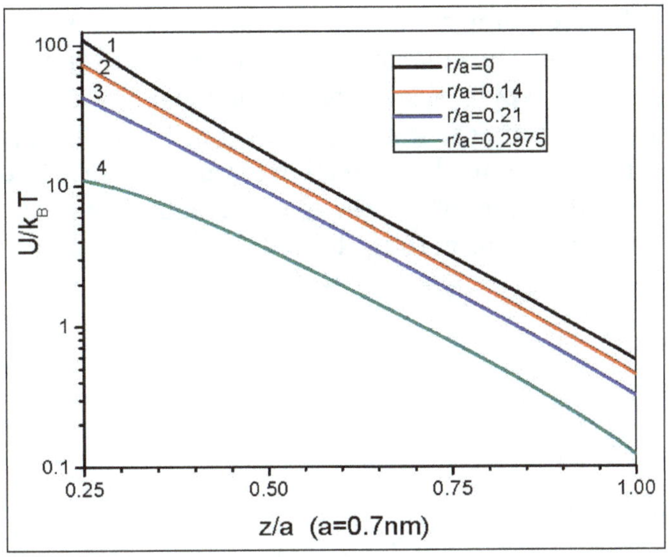

The parameter U/k_BT dependence on z. Curve 1 corresponds to r/a = 0, 2 – r/a = 0.14, 3 - r/a = 0.21, and 4 – r/a =0.30.

The value U/k_BT decreases exponentially with the distance from the membrane, wherein a distance z/a = 0.5, the ratio U/k_BT is about 10. That is still sufficiently deep potential well, but already at z/a =1, the influence of the membrane on the ions can be neglected.

An effective potential of the ion across the membrane can be determined. Let the ion be a sphere of radius r_i, and r_h is the distance from the positive charge of the dipole to the head of the ion.

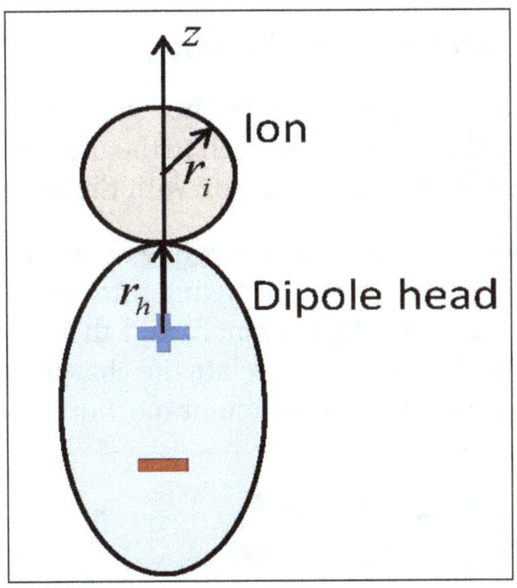

The negative ion located near the head of the phospholipid molecule dipole. Ion is a charged sphere of radius r_i, and r_h is the distance from the positive charge of the dipole to the head of the ion. In this case, the effective reduced ion potential,

$$\frac{U_{ef}}{k_BT} = \frac{2\pi}{k_BT}\frac{3}{4\pi r_i^3}\int_{r_h}^{r_h+2r_i} dz \int_0^{\sqrt{r_i^2-(r_h+r_i-z)^2}} U(r,z)\,r\,dr\,dz$$

The dependence of the effective potential of the ion Cl^- in the potential well on the size of the head hr is shown in figure above. The chlorine ion radius r_i = 0.18 nm, and the atomic weight is M_i = 35.4. It is seen that for the typical size of the dipole head r_h = 0.2 nm (r_h/a = 0.29) and the ion radius r_i = 0.18 nm, the reduced potential depth is $U_{ef}/k_BT \approx 10$. Thus, the trapped ion is quite firmly bound to the membrane. It should be noted that for electrolytes of a different composition of negative ions, for example in the axon, where the major negative ions are anion groups of macromolecules and phosphates, estimations of the potential will differ from that shown in figure below:

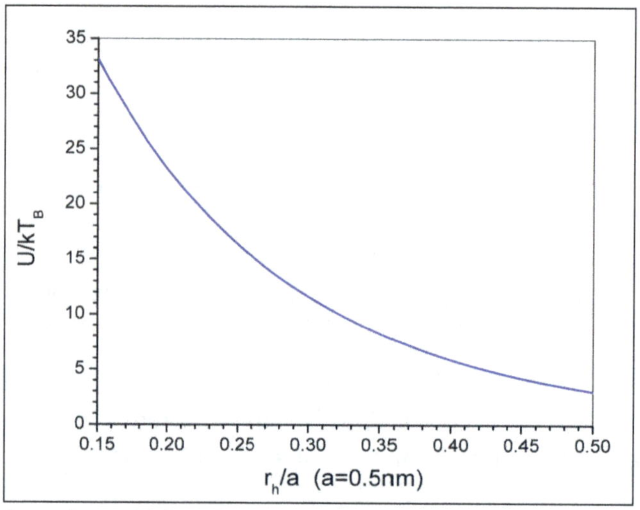

The dependence of the effective potential of the ion on the size of the head at the assumed radius of the ion r_i = 0 18 nm (Cl^-).

Evaluation of the Membrane Surface Charge

For simplicity, we assume the ions as a point charges, but will take into account the finite size of the ion as the minimal distance that the ion can approach to the membrane. As indicated above, this is a standard approach in models of the ions interaction with the dielectric surfaces.

We assume that the space charge on the membrane does not affect its structure, i.e. the dipole heads. Since the field of the membrane (without taking into account the ionic charge on it) at a distance of the order of mesh size is negligibly small, and the distance between the free ions in the solution is larger than the mesh size, we can relate the charge on the membrane surface to the charges in solution by equating the flows at the membrane/liquid boundary.

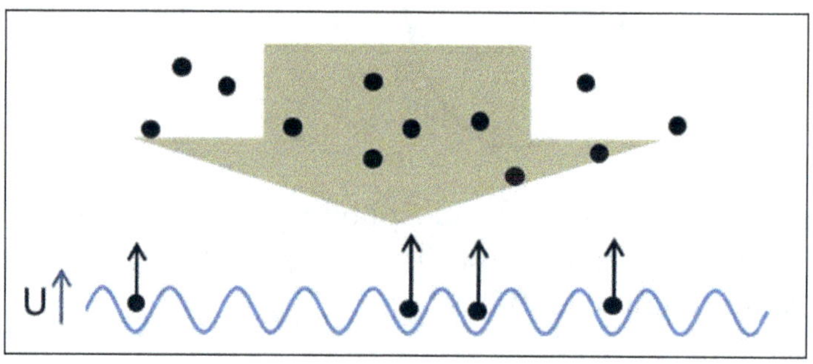

A scheme of the ions flows near the membrane. The ions flux to the membrane and "evaporation" of the ions from the potential wells of the depth U.

Let us estimate the flux of ions falling on the membrane. Assuming the Boltzmann distribution of ions outside the membrane:

$$n = n_\infty \exp\left(-\frac{q\varphi}{k_B T}\right)$$

where n_∞- is ion density at infinity. Assuming that the ions are in thermal equilibrium with the water molecules, we obtain the following estimate for the ion flux on the surface of the membrane:

$$P_{in} = \frac{1}{2\sqrt{\pi}} n_\infty v_{H_2O} \left(\frac{M_{H_2O}}{M_i}\right)^{1/2} \exp\left(-\frac{q\varphi}{k_B T}\right)$$

Here, $v_{H_2O}\left(\frac{M_{H_2O}}{M_i}\right)^{1/2}$ is the averaged ion velocities in solution, M_{H_2O}, M_i are the masses of the water molecule and the ion, correspondingly.

On the other hand, an estimate of the ion flux "evaporating" from the surface of the membrane is as follows. The flow of ions from the surface of the membrane can be considered as a process of evaporation of the liquid with a work function equal to the potential well U. The ion in the well is exposed to random impacts by water molecules, while the ion's presence time in the potential well τ_i can be estimated as:

$$\tau_i \approx \tau_{iH_2O}\left(\frac{M_i}{M_{H_2O}}\right)\left(\exp(U/kT)-1\right)$$

Here, $\tau_{iH_2O}\left(\frac{M_i}{M_{H_2O}}\right)$ is the characteristic time of energy exchange between the ion and the water molecules. In equation above we have taken into accounts that at U/kT <<1 the ions are reflected from the membrane, i.e., the time delay is zero. For estimates τ_{iH_2O} can be assumed equal to the collision time of water molecules in a solution i H H O T/v 2 0 2 τ, = λ, 9 0.2 10 2 − = · λH O nm is the mean free path of a molecule of water, and Tv is the thermal velocity of water molecules. Accordingly, the number of the ions leaving the unit surface of the $\tau_{iH_2O} = \lambda_{H_2O} / v_T, \lambda_{H_2O} = 0.2 \cdot 10^{-9}$ nm is the mean free path of a molecule of water, and v_T is the thermal velocity of water molecules. Accordingly, the number of the ions leaving the unit surface of the membrane per unit time is equal to:

$$P_{out} = \frac{1}{a^2}\frac{N_i}{\tau_i} = \frac{N_i v_{H_2O}}{a^2 \cdot \lambda_{H_2O}}\left(\frac{M_{H_2O}}{M_i}\right)\frac{1}{\exp(U/k_B T)-1}$$

Here $1/a^2$ is the number of potential wells per unit membrane surface; N_i is the relative population of the potential wells with ions. Equating the incident flux of the ions on the membrane to the "evaporating" flow from it, we obtain the relative population of the potential wells:

$$\frac{N_i}{1-N_i} = \frac{1}{2\sqrt{\pi}} n_\infty a^2 \cdot \lambda_{H_2O}\left(\frac{M_i}{M_{H_2O}}\right)^{1/2} \exp\left(-\frac{e\varphi_0}{k_B t}\right)\left(\exp\left(\frac{U}{k_B T}\right)-1\right)$$

In $P_{out} = \frac{1}{a^2}\frac{N_i}{\tau_i} = \frac{N_i v_{H_2O}}{a^2 \cdot \lambda_{H_2O}}\left(\frac{M_{H_2O}}{M_i}\right)\frac{1}{\exp(U/k_BT)-1}$ we took into account that the probability of the incident ion being captured for in a potential well is proportional to $1-N_i$.

Now we will find the distribution of electric potential in the liquid. Following the Debye approximation for the liquid electrolyte:

$$\varphi = \varphi_o \exp(-x/\lambda_D)$$

Where $\lambda_D = \sqrt{\frac{\varepsilon\varepsilon_o k_B T}{2n_\infty q^2}}$ is the Debye length.

The field on the membrane surface:

$$\left.\frac{\partial\varphi}{\partial x}\right|_{x=0} = \frac{\sigma}{2\varepsilon\varepsilon_0}$$

Where σ is the surface charge density. For the case when the upper and lower surfaces of the membrane are equally charged $\sigma = 2N_i q/a^2$, we obtain the potential at the membrane surface:

$$\varphi_o = \frac{\lambda_D}{\varepsilon\varepsilon_o}\frac{q \cdot N_i}{a^2} = \sqrt{\frac{k_B T}{2\varepsilon\varepsilon_0 n_\infty}} \cdot \frac{N_i}{a^2}$$

Substituting equation above into $\frac{N_i}{1-N_i} = \frac{1}{2\sqrt{\pi}} n_\infty a^2 \cdot \lambda_{H_2O}\left(\frac{M_i}{M_{H_2O}}\right)^{1/2} \exp\left(-\frac{e\varphi_o}{k_B t}\right)\left(\exp\left(\frac{U}{k_B T}\right)-1\right)$, we obtain the equation for the value of N_i:

$$\frac{N_i}{1-N_i} = \frac{1}{2\sqrt{\pi}} n_\infty a^2 \cdot \lambda_{H_2O}\left(\frac{M_i}{M_{H_2O}}\right)\exp(-\eta N_i) \cdot \left(\exp\left(\frac{U}{k_B T}\right)-1\right)$$

$$\eta = \frac{q}{a^2}\sqrt{\frac{1}{2\varepsilon\varepsilon_0 n_{\infty k_B T}}}$$

For parameter values that are typical for a physiological solution: $n_\infty = 0.9 \cdot 10^{26}$ m^{-3}, $a^2 = 0.5 \cdot 10^{-18}$ m^2, T = 300 K, $\lambda_{H2O} = 0.2$ nm, for the chloride ion $(M_i / M_{H_2O})^{1/2} \approx 1.41$ we obtain from above equation $\eta \approx 13.9$, and, therefore,

$$\frac{N_i}{1-N_i} = 0.025 \cdot (\exp(U/k_B T)-1) \cdot \exp(-13 \cdot N_i)$$

The calculated dependence of N_i on the reduced potential $\frac{U}{k_B T}$ is shown in Figure below.

Dependencies of the relative occupancy of the membrane N_i and the surface charge density σ on the reduced potential U/k_BT. Curve 1 corresponds to $n_\infty = 5 \cdot 10^{26}$, 2 - $n_\infty = 10^{26}$, 3 - $n_\infty = 5 \cdot 10^{25}$, 4 - $n_\infty = 10^{25}$, 5 - $n_\infty = 10^{25}$, 5 - $n_\infty = 5 \cdot 10^{24}$, 6 - $n_\infty = 10^{24}$.

The maximum surface charge density in Figure above corresponds to the case where all of the cells are full, i.e. one negative ion is in every cell. It is seen that, even for small values U/k_BT, the relative occupancy of the potential wells with charges is large enough and the charge density of the charges, which are tightly bounded to the membrane, can reach hundredths of coulomb per square meter. It should be noted that in this estimation we did not consider the interaction between the ions bonded to the membrane, which can certainly play an important role for a more accurate calculation of the occupancy of the potential wells.

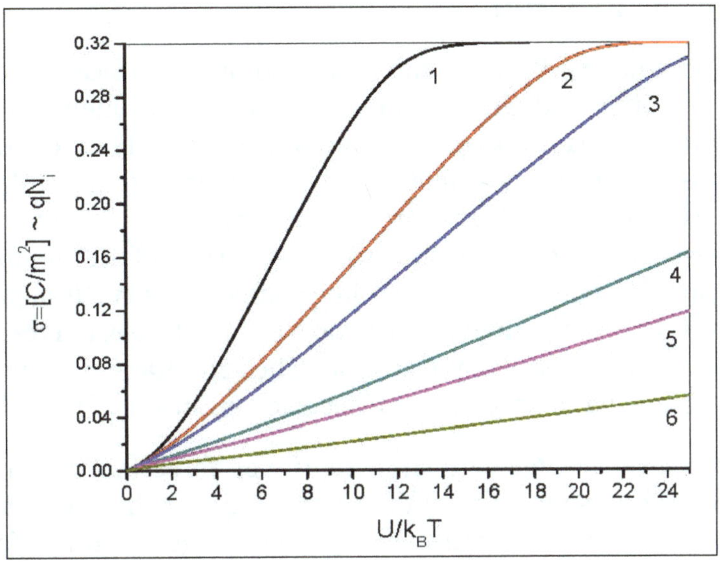

Let us estimate the charge of the free ions of one sign in the non-quasineutral region in the vicinity of the membrane surface.

$$q_{free} \approx q \cdot n_{\infty} \cdot r_D \approx 0.013 C / m^2$$

For typical values of the potential $U/k_BT \approx 5-15$, which correspond to the size of the head $r_h \sim 0.2 - 0.3$ nm, a bounded charge sitting on the membrane surface is much larger than the charge associated with unbounded ions.

Membrane and Transmembranes Potential

An electrical potential difference exists between two locations when there is a net separation of charge between the two locations. All living cells maintain a potential difference across their membrane. This membrane potential is due to difference in concentration and permeability of sodium and potassium on either side of the membrane. Because of the unequal concentrations of ions across a membrane, the membrane has an electrical charge, inside negative.

Changes in membrane potential to elicit action potentials and give cells the ability to send messages to different parts of the body. Some cells, such as nerve and muscle cells are capable of generating quickly changing electrochemical impulses at their membranes, and these impulses are used to transmit signals along the nerve or muscle membranes. In other types of cells, such as macrophages, glandular cells, and ciliated cells, local changes in membrane potentials also stimulate many of the cellular functions. The membrane potential is created due to various proteins embedded within the cell membrane, as well as the structure of the phospholipid lipid bilayer. From a physiological point of view, the membrane potential is responsible for sending messages to and from the central nervous system. It is also very important in cellular biology and shows how cell biology is fundamentally connected with electrochemistry and physiology.

Physiology of Membrane Potential

Cell membrane works as a barrier between the inner and outer surface of a cell. In discussing the concept of membrane potentials and how they function, the establishment of membrane potential is essential. The membrane that surrounds a cell is composed of phospholipids, glycolipids, cholesterol and proteins. The primary structure is a lipid bilayer. Phospholipid molecules have an electrically charged head that attracts water and a hydrocarbon tail that repels water; they line up side by side in two opposing layers, with their heads on the inner or outer surface of the membrane and their tails in the core, from which water is excluded. The other lipids affect the structural properties of the membrane. The proteins embedded in the lipid bilayer, the Na^+/K^+ ATPase, ion transporters, and voltage gated channels, and it is the site of vesicular transport. The membrane structure regulates entry ions and exit of ions to determine the concentration of specific ions inside of the cell.

Membrane potentials in cells are determined primarily by three factors: a) the concentration of ions on the either side of the membrane; b) the permeability of the cell membrane to those ions through specific ion channels; and c) by the activity of electrogenic pumps (e.g., Na^+/K^+ - ATPase and Ca^{2+} transport pumps) that maintain the ion concentrations across the membrane.

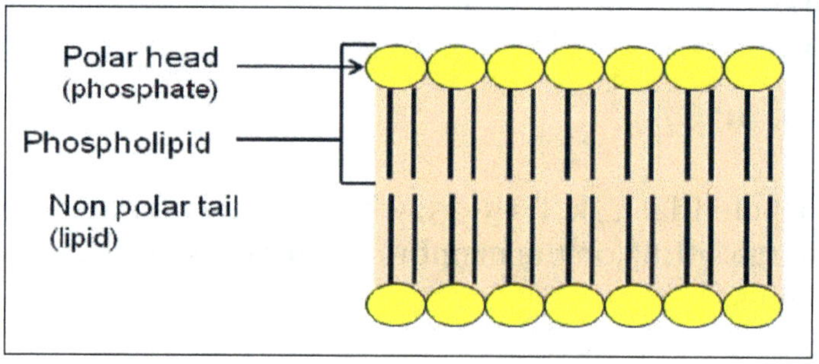

Model of membrane lipid bilayer.

Understanding Membrane Potential

The following points important to understand how the membrane potential works. The difference between the electrical and chemical gradient is important.

- Electrical Gradient:
  - Opposes the chemical gradient.
  - Represents the difference in electrical charge across the membrane.
- Chemical Gradient:
  - Opposes the electrical gradient.
  - Represents the difference in the concentration of a specific ion across the membrane.

- A good example is K^+. The membrane is very permeable to K^+ and the K^+ ion concentration inside the cell is high, therefore a positive charge is flowing out of the cell along with K^+. The K^+ ion concentration inside the cell decreases causing the concentration gradient to flow towards the outside of the cell. This also causes the inside of the cell to become more electronegative increasing its electrical gradient.
- The Nernst equation can help us relate the numerical values of concentration to the electrical gradient.
- Leak Channels:
  - Channels that are always open.
  - Permit unregulated flow of ions down an electrochemical gradient.
- Na^+/K^+ ATPase:
  - Actively transports Na^+ ions out of the cell and K^+ ions into the cell.
  - Helps to maintain the concentration gradient and to counteract the leak channels.

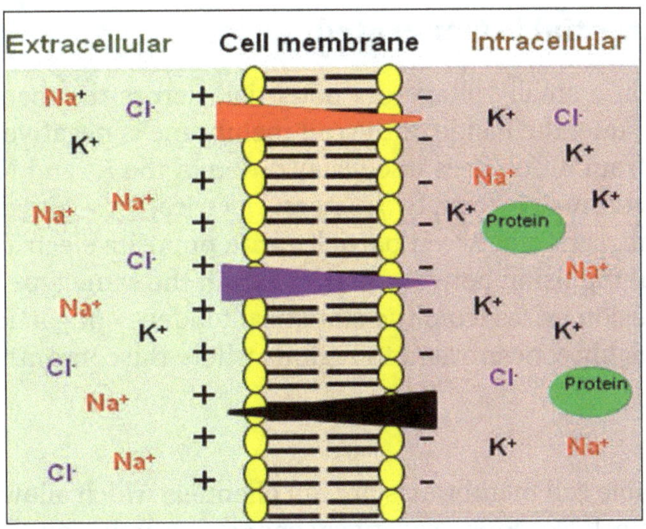

Differences in the concentration of ions on both sides of a cell membrane produce a voltage difference called the membrane potential. The major contributions usually come from sodium (Na^+) and chloride (Cl^-) ions which have high concentrations in the extracellular side, and potassium (K^+) ions, which along with large protein anions have high concentrations in the intracellular side of the membrane.

How to Calculate Membrane Potential of a Cell?

The calculation for the charge of an ion across a membrane, The Nernst Potential, is relatively easy to calculate. The equation is as follows:

$$(RT/zF) \log([I]_{out}/[I]_{in})$$

RT/F is approximately 61, therefore the equation can be written as:

$$(61/z) \ln([I]_{out}/[I]_{in})$$

Derived under resting membrane conditions when the work required to move an ion across the membrane (up its concentration gradient) equals the electrical work required to move an ion against a voltage gradient.

- R= Gas constant (8.314 jules/°K.mol).
- T= Temperature (°K).
- F= Faraday constant (96, 000 coulombs/mol).
- z= The electric charge on the ion. For example, z is +1 for K^+, +2 for Mg^{2+} etc. z have no unit.
- $[I]^{out}$= Ion concentration outside the cell.
- $[I]^{in}$= Ion concentration inside the cell.

The only difference in the Goldman-Hodgkin-Katz equation is that is adds together the concentrations of all permeable ions as follows:

$$(RT/zF)\ \log([K^+]_o+[Na^+]_o+[Cl^-]_o\ /[K^+]_i+[Na^+]_i+[Cl^-]^i)$$

How Membrane Potential is Generated

Cell membranes maintain a small voltage or "potential" across the membrane in its normal or resting state. In the rest state, the inside of the cell membrane is negative with respect to the outside. The voltage arises from differences in concentration of the K^+ and Na^+ ions. The membrane potential of a cell derives ultimately from two factors: electrical force and diffusion. Electrical force arises from the mutual attraction between particles with opposite electrical charges (positive and negative) and the mutual repulsion between particles with the same type of charge (both positive and both negative). Diffusion arises from the statistical tendency of particles to redistribute from regions where they are highly concentrated to regions where the concentration is low.

Electrical Force

In the selectively permeable cell membranes are ion channels which allow K^+ ions to pass into the interior of the cell, but block Na^+ ions. Negatively charged proteins on the interior of the cell are also denied passage across the membrane. In addition, there are active transport mechanisms at work. There is a process called the Na^+/K^+ ATPase, which utilizes ATP to pump out three Na^+ ions and pump in two K^+ ions. The collective action of these mechanisms leaves the interior of the membrane about -70 mV with respect to the outside.

Membrane Potentials Caused by Diffusion

"Diffusion Potential" caused by an ion centration difference on the two sides of the membrane. The potassium concentration is great inside, but very low outside the membrane. Let us assume that the membrane in this instance is permeable to the potassium ions, but not to any other ions. Because of the large potassium concentration gradient from inside to outside, there is a strong tendency for extra numbers of potassium ions to diffuse outward through the membrane. As they do so, they carry positive electrical charges to the outside, thus creating electro positivity outside the membrane and electro negativity inside because of negative anions that remain behind and do not diffuse outward with the potassium. Within a millisecond or so, the potential difference between the inside and outside, called the diffusion potential, becomes great enough to block further net potassium diffusion to the exterior, despite the high potassium ion concentration gradient. The same phenomenon occurs for sodium ions, but this time with high concentration of sodium ions outside the membrane and low sodium inside. These ions are also positively charged. This time, the membrane is highly permeable to the sodium ions, but impermeable to all other ions. Diffusion of the positively charged sodium ions to the inside creates a membrane potential of opposite polarity with negativity outside and positivity inside. Again, the membrane potential rises high enough within milliseconds to block further net diffusion of sodium ions to the inside. A concentration difference of ions across a selectively permeable membrane can, under appropriate conditions, create a membrane potential. The rapid changes in membrane potentials observed during nerve and muscle impulse transmission result from the occurrence of such rapidly changing diffusion potentials.

The Contribution of the Na^+/K^+ ATPase

The Na^+/K^+ ATPase provides an additional contribution to the resting potential. Na^+/K^+ ATPase pumps continuously three sodium ions to the outside for each two potassium ions pumped to the inside of the membrane. The fact that more sodium ions are being pumped to the outside than potassium to the inside causes continual loss of positive charges from inside the membrane; this creates an additional degree of negativity on the inside beyond that which can be accounted for by diffusion alone. Therefore, the net membrane potential with all these factors operating at the same time is about −90 millivolts. In summary, the diffusion potentials alone caused by potassium and sodium diffusion would give a membrane potential of about −86 millivolts, almost all of this being determined by potassium diffusion. Then, an additional −4 millivolts contribute to the membrane potential by the continuously acting electrogenic Na^+/K^+ ATPase, giving a net membrane potential of −90 millivolts. If the cell membranes were simply permeable to these ions, they would approach an equilibrium with equal concentrations on each side of the membrane, and hence no voltage difference.

Transmembrane Potential

This is the difference in electrical potential across the cell membrane. There are a number of factors which contribute to this difference:

- Differences in ion concentrations across the membrane:
  - Sodium ion concentration is high outside cell.
  - Potassium ion concentration is high inside cell.
- The cell membrane selective permeability:
  - K^+ ions are more permeable than Na^+ ions.
  - Proteins are too large to cross the membrane.
- As a result of the above factors, the cytoplasm tends to be negatively charged. The extracellular side is positively charged. The resistance of the lipid bilayer to a free flow of substances across the membrane helps to maintain this potential difference across the membrane. The typical values of membrane potential range from −40 mV to −90 mV.

Resting Potential

The inside of the cell has a negative voltage with respect to the outside of the cell. Under resting conditions, this is called the resting membrane potential. A resting cell tends to show a fairly stable transmembrane membrane potential across the membrane. With appropriate stimulation of the cell, this negative membrane potential may transiently become positive owing to the generation of an action potential.

Resting membrane potential varies according to type of cells. Excitable cells; nerves and muscle cells have higher potentials than other cells (epithelial cells and connective tissue cells). For example:

Cell Type	Resting Potential
Adipose cell	-40 mV
Thyroid cell	-50 mV
Neuron	-70 mV
Skeletal muscle cell	-85 mV
Cardiac muscle cell	-90 mV
Smooth muscle cell	-50 mV
Astrocytes	-80/-90 mV
Erythrocytes	-12 mV

Action Potential

The action potential is generated by the activation (opening) and subsequent inactivation of voltage-gated sodium channels and, with a slight delay, the opening of voltage-gated potassium channels. In mammals there are 10 voltage-gated sodium channels, and many of these are expressed in neurons and localized in axons. In addition, there are at least 20 voltage-gated potassium channels in vertebrates. Sodium–potassium ATPase although the immediate basis of the action potential is the activation of voltage-gated sodium and potassium channels, long-term support of the action potential requires sodium–potassium ATPase to maintain the concentration gradients. It has been often emphasized that if the sodium ATPase is blocked (e.g., with ouabain) in a squid axon, hundreds of thousands of action potentials can still be generated because few sodium and potassium ions cross the membrane for each action potential.

The bit of information in nerves is the action potential, a fast electrical transient in the transmembrane voltage that propagates along the nerve fiber. In the resting state, the membrane potential of the nerve fiber is about -60 mV. When the action potential is initiated the membrane potential becomes less negative and even reverses sign (overshoot) within 1 ms and then goes back to the resting value in about 2 ms, frequently after becoming even more negative than the resting potential. The generation of the rising phase of the action potential was explained by a conductance to Na ions that increases as the membrane potential is made more positive. This is because, as the driving force for the permeating Na ions was in the inward direction, more Na ions come into the nerve and makes the membrane more positive initiating a positive feedback that depolarizes the membrane even more. This positive feedback gets interrupted by the delayed opening of another voltage dependent conductance that is K selective. The driving force for K ions is in the opposite direction of Na ions; thus, K outward flow repolarizes the membrane to its initial value.

Measurement of Transmembrane Potential by Microelectrodes

If a voltmeter is attached to the two terminals of a battery, a voltage difference will be measured across the two terminals. Measurements of membrane potentials with glass microelectrodes face some difficulties when delicate cells are the subject of the impalement. Delicate cells can be those in which the membrane does not seal properly around the electrode glass, they may be small like lymphocytes where the microelectrode tip is relatively too large, bigger like sea urchin eggs but easily activated by electrode impalement. The membrane potential is measured with glass microelectrodes filled with solutions which conduct charge. The microelectrode is inserted through the membrane into the cell. The voltmeter measures the difference in electrical charge between two points. The potential difference is measured in millivolts (mV). The resting membrane potential is measured on a relative scale.

The reference electrode is placed in the extracellular fluid. The extracellular fluid is designated as the ground and assigned a charge of 0 mV. In reality, the extracellular fluid is not neutral and has an excess of positive charge that balances the excess of negative charge inside the cell. The numerical value of the membrane potential is generally negative, meaning that the inside of the cell is negative with respect to the outside solution, which is taken as the reference or zero value.

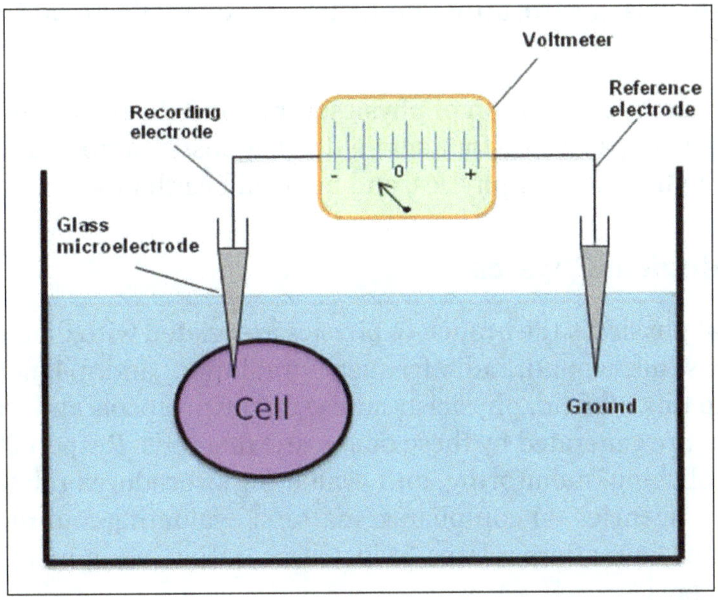

References

- Reading-structure-of-the-cell-membrane: lumenlearning.com, Retrieved 08, July 2020
- The-surface-charge-of-a-cell-lipid-membrane: arxiv.org, Retrieved 22, April 2020

4

Medical Biophysics

Medical biophysics is the branch of biophysics that makes use of physics to describe biological processes for medical applications. Some of the significant topics studied under medical biophysics are electromyography, ultrasonography, blood flow measurement, cardiac pacemakers, flow cytometry, etc. This chapter discusses in detail these significant aspects related to medical biophysics.

Medical physics is the use of physics principles in the practice of medicine. It is most often used to describe physics applications related to the use of radiation in medicine—for example, the physics of diagnostic radiology, radiation oncology, and nuclear medicine. More broadly defined, medical physics may include the physics of other electromagnetic waveforms used in medical procedures such as electrocardiography (the study of electrical impulses in the heart) and laser surgery.

Medical physics refers to the application of physics in medical diagnosis and treatment. The bulk of medical physics is encompassed by four subfields: diagnostic radiological physics, therapeutic radiological physics, medical nuclear physics, and medical health physics.

Diagnostic Radiological Physics

Diagnostic radiological physics is the branch of physics associated with diagnostic procedures that use x-rays, gamma rays, ultrasound, radio frequency radiation, and magnetic sources (magnetic resonance imaging). In this subfield, physicists advise on the protocols and technology used for the creation of images that are generated by these diagnostic methods. Responsibilities of the medical physicist include establishing, monitoring, and evaluating procedures related to equipment use; reporting to regulatory agencies on compliance matters; evaluating and monitoring equipment; and acting as consultant on matters related to instrumentation, equipment, and use of these radiological imaging systems.

Therapeutic Radiological Physics

Therapeutic radiological physics concerns itself with the physics of therapeutic procedures that use x rays, gamma rays, neutrons, charged particles, and radionuclides from sealed sources (radioactive material that is sealed permanently in a container). These therapeutic procedures are often used in the treatment of cancer and include external beam therapy (where ionizing radiation is directed at the cancer site) and brachytherapy (where containers with radioactive material are placed near or in the tumor). Duties of the therapeutic radiological physicist include providing consultation on matters related to appropriate radiation dose and risks to patients; managing procedures and equipment related to dose and delivery of therapeutic radiation; reporting to regulatory agencies on compliance matters; and designing, evaluating, and monitoring radiation safety program related to therapeutic radiological procedures.

Medical Nuclear Physics

Also known as nuclear medicine physics, medical nuclear physics is the study of physics related to medical procedures requiring the use of radionuclides (except those radionuclides from sealed sources). These procedures may be diagnostic or therapeutic, and include such procedures as single photon emission computed tomography (SPECT), positron emission tomography (PET), and radio-immunotherapy (radioisotopes attached to molecules that can be targeted to cancer cells). The medical nuclear physicist acts as consultant on matters related to appropriate radionuclide dose and risks; manages procedures and equipment related to dose and delivery of radionuclide imaging equipment; reports to regulatory agencies on compliance matters; and designs, evaluates, and monitors radiation safety program related to the nuclear medicine facility.

Medical Health Physics

The medical health physicist specializes in issues related to radiation safety in medical procedures. There is some overlap between medical health physics and the three other subfields since the use of radiation for medical purposes always requires some safety safeguards. The medical health physicist takes part in designing and specifying the radiation shielding required to protect patients, health care workers, and the general public; conducts risk assessment of procedures and protective equipment used in radiological and nuclear medicine; acts as consultant on issues related to radiation safety in a medical context; and evaluates and monitors compliance with regulatory radiation guidelines.

Basics of Instrumentation in Medical Physics

Measurement is an important subsystem of a mechatronics system. Its main function is to collect the information on system status and to feed it to the micro-processor(s) for controlling the whole system. Measurement system comprises of Sensors, Transducers Signal Processing Devices.

A transducer is a device that converts one form of energy to another form of energy. Energy types include/(but are not limited to) electrical, mechanical, electromagnetic (including light), chemical, acoustic, and thermal energy.

INPUT ENERGY → TRANSDUCER → OUTPUT ENERGY (SIGNAL)

SENSING PROCESS

The word "Transducer" is the collective term used for both Sensors which can be used to sense a wide range of different energy forms such as movement, electrical signals, radiant energy, thermal or magnetic energy etc. Transducer contains two parts that are closely related to each other:

- Sensing element is called sensor: A device producing measureable response to Change in physical conditions such as temperature, pressure etc.

- Transduction element converts the sensor output to suitable electrical form.

Parameters → Sensing element →Transduction element →Electrical Signal
such as
temperature,
pressure

SENSOR OUTPUT

Classification of Transducers

- Analog and digital transducers.
- Primary and Secondary transducers.
- Transducers and Inverse transducers.
- On the basis of transduction principle used.
- Active and passive transducers.

Analog and Digital Transducers

There are many different types of sensors and transducers:

- Input and Output type.
- Analogue and Digital type.

Input and Output Type Transducers

The type of input or output transducer being used, really depends upon the type of signal or process being "Sensed" or "Controlled" but a sensor and transducers can be described as devices that converts one physical quantity into another.

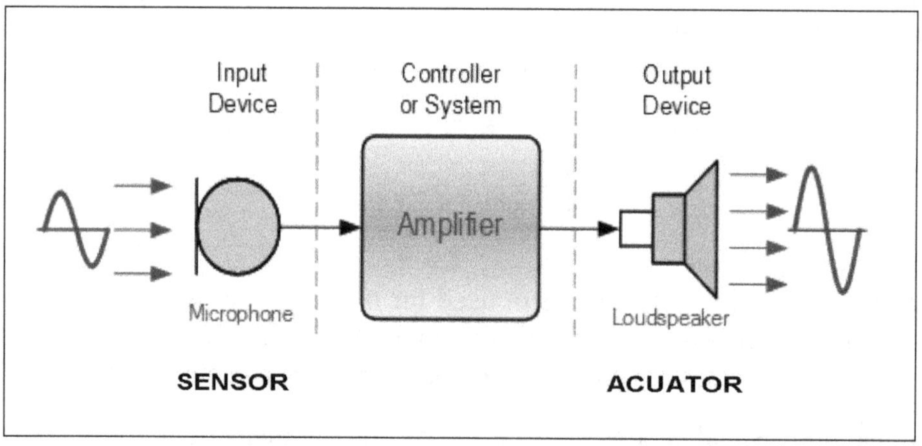

Input type transducers or sensors, produce a voltage or signal output response which is proportional to the change in the quantity that they are measuring (the stimulus). The type or amount of the output signal depends upon the type of sensor being used. But generally, all types of sensors can be classed as two kinds, either Passive Sensors or Active Sensors.

An actuator is a transducer that accepts energy and produces the kinetic energy of movement (action). The energy supplied to an actuator might be electrical or mechanical (pneumatic, hydraulic, etc). An electric motor and a hydraulic cylinder are both actuators converting electrical energy and fluid power into motion for different purposes.

Combination transducers have both functions; they both detect and create action. The most common example is an antenna, a transducer of radio waves that can transmit, receive, or both (transceiver). Another example is the typical ultrasonic transducer, which switches back and forth many times a second between acting as an actuator to produce ultrasonic waves, and acting as a sensor to detect ultrasonic waves. Rotating a DC electric motor's rotor will produce electricity, and voice-coil speakers can also act as microphones.

Quantity being Measured	Input Device (Sensor)	Output Device (Actuator)
Light Level	Light Dependant Resistor (LDR) Photodiode Photo-transistor Solar Cell	Lights & Lamps LED's & Displays Fibre Optics
Temperature	Thermocouple Thermistor Thermostat Resistive Temperature Detectors	Heater Fan
Force/Pressure	Strain Gauge Pressure Switch Load Cells	Lifts & Jacks Electromagnet Vibration
Position	Potentiometer Encoders Reflective/Slotted Opto-switch LVDT	Motor Solenoid Panel Meters
Speed	Tacho-generator Reflective/Slotted Opto-coupler Doppler Effect Sensors	AC and DC Motors Stepper Motor Brake
Sound	Carbon Microphone Piezo-electric Crystal	Bell Buzzer Loudspeaker

Analogue and Digital Type Sensors

Analogue Sensors

Analogue Sensors produce a continuous output signal or voltage which is generally proportional to the quantity being measured. Physical quantities such as Temperature, Speed, Pressure, Displacement, Strain etc are all analogue quantities as they tend to be continuous in nature. For example, the temperature of a liquid can be measured using a thermometer or thermocouple which continuously responds to temperature changes as the liquid is heated up or cooled down.

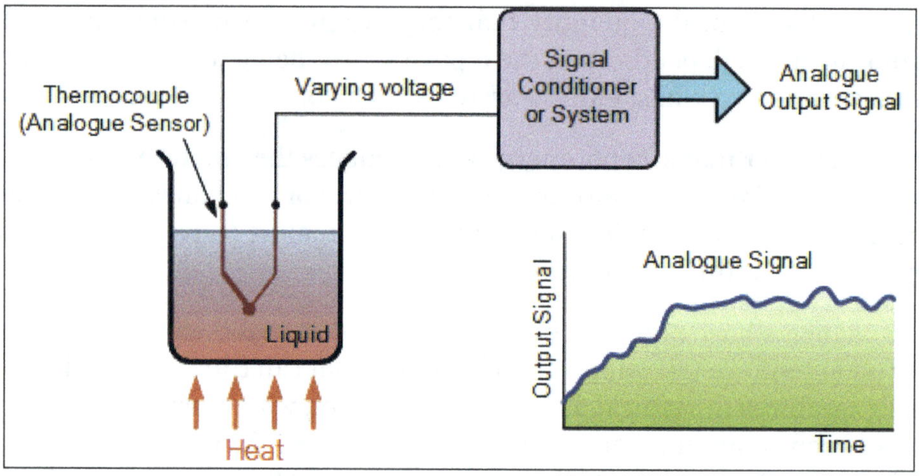

Thermocouple to Produce an Analogue Signal

Analogue sensors tend to produce output signals that are changing smoothly and continuously over time. These signals tend to be very small in value from a few micovolts (μV) to several milli-volts (mV), so some form of amplification is required. Then circuits which measure analogue signals usually have a slow response and/or low accuracy. Also analogue signals can be easily converted into digital type signals for use in micro-controller systems by the use of analogue-to-digital converters, or ADC's.

Digital Sensors

Digital Sensors produce a discrete digital output signals or voltages that are a digital representation of the quantity being measured. Digital sensors produce a Binary output signal in the form of a logic 'I' or a logic 'o', ("ON" or "OFF"). This means then that a digital signal only produces discrete (non-continuous) values which may be outputted as a single 'bit', (serial transmission) or by combining the bits to produce a single 'byte' output (parallel transmission).

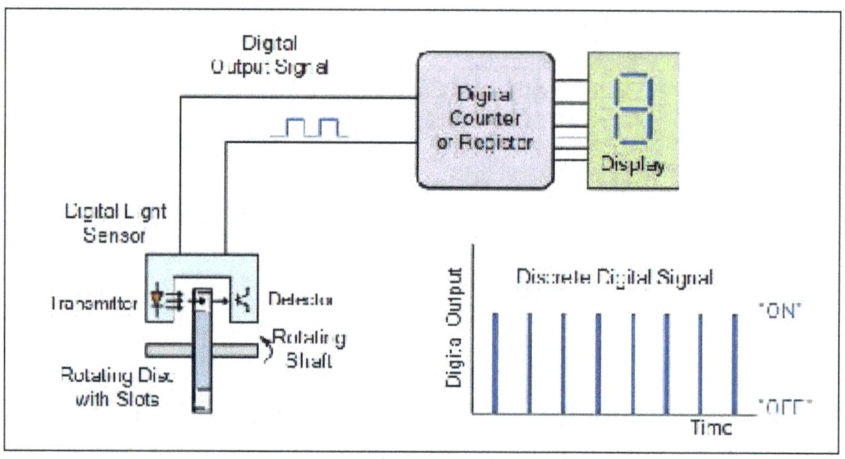

Light Sensor to Produce a Digital Signal

The speed of the rotating shaft is measured by using a digital LED/Opto-detector sensor. The disc

is fixed to a rotating shaft (for example, from a motor or robot wheels), has a number of transparent slots within its design. As the disc rotates with the speed of the shaft, each slot passes by the sensor in turn producing an output pulse representing logic 'I' or logic 'o' level.

Transducers

On the basis of methods of applications, transducers may be classified into Primary and Secondary transducer.

Primary Transducers

- When the input signal is directly sensed by the transducers and physical phenomena is converted into the electrical form directly then such a transducer is called the primary transducer. Example: Thermistor.
- Thermistor senses the temperature directly and causes the change in resistor with the change in temperature.

Secondary Transducers

- When the input signal is sensed first by some detector or sensor and then its output being of some form other than input signal is given as input to a transducer for conversion into electrical form, then such a transducer is referred to as Secondary Transducer.
- Example: In case of Pressure measurement, burden tube is a Primary sensor which converts pressure first into displacement, then the displacement is converted into an output voltage by an LVDT.
- LVDT is a secondary transducer.

Inverse Transducers

- It is a devise which converts an electrical quantity into a non-electrical quantity.
- Example: Piezoelectric crystal.

On the Basis of Transduction Principle used:

- Resistive Transducers,
- Capacitive Transducers,
- Inductive Transducers,
- Voltage and Current Generating Transducers.

Active and Passive Transducers

On the basis of methods of Energy Conversion used, transducers are classified as Active and Passive Transducers.

- Active Transducers: Self generating type transducers which develop their output in the form of electrical voltage or current without any auxiliary source. Examples- Thermocouples, Piezoelectric crystals.

- Passive Transducers: Transducers in which electrical parameters such as resistance, inductance or capacitance changes with the change in input signal. These types of transducers require external power source for energy conversion. Example – Resistive inductive transducers. In these types of transducers, electrical parameters cause a change in voltage, current or frequency of the external power source.

In these types of transducers, electrical parameters cause a change in voltage, current or frequency of the external power source.

Applications

Electromagnetic

Transducers are used in electronic communications systems to convert signals of various physical forms to electronic signals, and vice versa.

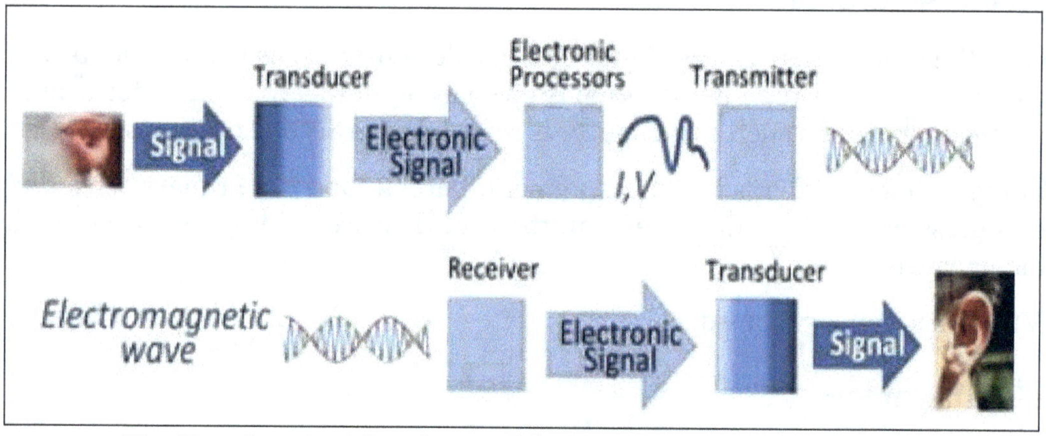

First Transducer is a Microphone and the Second Transducer is a Speaker.

- Antenna: Converts propagating electromagnetic waves to and from conducted electrical signals.
- Magnetic cartridge: Converts relative physical motion to and from electrical signals.
- Tape head, Disk 'Read and Write': Converts magnetic fields on a magnetic medium to and from electrical signals.
- Hall effect sensor: Converts a magnetic field level into an electrical signal.

Electrochemical

- pH probes.
- Electro-galvanic fuel cell.
- Hydrogen sensor.

Electromechanical

Electromechanical output devices are generically called actuators.

- Electroactive polymers.
- Galvanometer.
- Microelectromechanical systems.
- Rotary motor, linear motor.
- Vibration powered generator.
- Potentiometer when used for measuring position.
- Linear variable differential transformer or Rotary variable differential transformer.
- Load cell: Converts force to mV/V electrical signal using strain gauge.
- Accelerometer.
- Strain gauge.
- String potentiometer.
- Air flow sensor.
- Tactile sensor.

Electroacoustic

- Loudspeaker, earphone: Converts electrical signals into sound (amplified signal → magnetic field → motion → air pressure).
- Microphone: Converts sound into an electrical signal (air pressure → motion of conductor/coil → magnetic field → electrical signal).
- Pickup (music technology): Converts motion of metal strings into an electrical signal (magnetism → electrical signal).
- Tactile transducer: Converts electrical signal into vibration (electrical signal → vibration).
- Piezoelectric crystal: Converts deformations of solid-state crystals (vibrations) to and from electrical signals.
- Geophone: Converts a ground movement (displacement) into voltage (vibrations → motion of conductor/coil → magnetic field → signal).
- Gramophone pickup – (air pressure → motion → magnetic field → electrical signal).
- Hydrophone: Converts changes in water pressure into an electrical signal.
- Sonar transponder (water pressure → motion of conductor/coil → magnetic field → electrical signal).

- Ultrasonic transceiver, transmitting ultrasound (transduced from electricity) as well as receiving it after sound reflection from target objects, availing for imaging of those objects.

Electro-Optical (Photoelectric)

- Fluorescent lamp: Converts electrical power into incoherent light.
- Incandescent lamp: Converts electrical power into incoherent light.
- Light-emitting diode: Converts electrical power into incoherent light.
- Laser Diode: Converts electrical power into coherent light.
- Photodiode, photoresistor, phototransistor, photomultiplier: Converts changing light levels into electrical signals.
- Photodetector or photoresistor or light dependent resistor (LDR): Converts changes in light levels into changes in electrical resistance.
- Cathode ray tube (CRT): Converts electrical signals into visual signals.

Electrostatic

- Electrometer.

Thermoelectric

- Resistance temperature detector (RTD): Converts temperature into an electrical resistance signal.
- Thermocouple: Converts relative temperatures of metallic junctions to electrical voltage.
- Thermistor (includes PTC resistor and NTC resistor).

Radioacoustic

- Geiger–Müller tube: Converts incident ionizing radiation to an electrical impulse signal.
- Receiver (radio).
- Transmitter-propagates electromagnetic transmissions to sound.

Electromyography

Electromyography is the graphing and study of the electrical characteristics of muscles. Resting muscle is normally electrically silent. However, when it is active, as during contraction or stimulation, an electrical current is generated, and the successive action potentials (impulses) can be registered on a cathode-ray oscilloscope screen in the form of continuous wavelike tracings. The visual recording, called an electromyogram, or EMG, is customarily accompanied by auditory monitoring.

Electromyography is used primarily to differentiate diseases of the muscles from diseases of the peripheral nerves. A needle electrode is inserted into the muscle, and the electrical activity of the muscle is measured. Records of muscle electrical activity are usually obtained during muscle relaxation, during voluntary contraction when needle electrodes are inserted into the muscle under study, and during muscle activity evoked by the stimulation of its nerve. The response to electrical stimulation allows the physician to determine the origin of muscle weakness.

Weakness or wasting of muscle is generally caused either by impairment of the nerves supplying it (neuropathic disorders, such as amyotrophic lateral sclerosis and polio) or by intrinsic muscle impairment or primary muscle disease (myopathy). In neuropathic disorders, there is usually increased spontaneous activity during muscle relaxation (fibrillation and fasciculation), together with reduced, altered, or absent normal muscle action potentials. In myopathies, there is frequently a reduction in the amplitude or duration of the muscle action potentials and an increase in the complexity of their wave form.

Physiology of Muscle Activation and Force Generation

Concept of Motor Unit

A motor unit consists of a motor neuron and the muscle fibers innervated by the axon. It's the functional unit of the neuromuscular system. Let's understand each component of motor unit separately. The group of motor neurons that innervate a single muscle is called motor nucleus or the motor neuron pool. There are two types of motor neurons. 1) The alpha motor neurons. 2) The gamma motor neurons. Both the classes of the motor neurons are unique in their on ways. The alpha motor neurons are like the alpha males, big and innervate lots of extra fusal muscle fibers, whereas the gamma motor neurons are smaller than the alpha neurons, innervating mainly the intrafusal muscle fibers. From the motor neuron pool all these neurons come out from the ventral column of spinal cord. So when an action potential reaches the motor neuron, it passes down to the different muscle fibers and activates them as well.

In a motor unit, the nerve synapse with the muscle fiber. The terminal end of the neuron is the presynaptic part, then comes the synaptic cleft which ends with the post synaptic membrane of the muscle fiber. The terminal ends of the nerve (i.e. presynaptic membrane) have various especial densities called the synaptic bouton. This specialized location mainly contains neurotransmitters. The action potential when arrive at the presynaptic membrane, these neurotransmitters are released and they diffuse through the synaptic cleft and attaches with the post synaptic membrane thereby producing end plate potentials. These excitatory end plate potentials summate further and develop the action potential. This action potential then brings out the muscle contraction.

Skeletal Muscle

The musculoskeletal system comprises one of the major tissue/organ systems in the body. The three main types of muscle tissue are skeletal, cardiac and smooth muscle groups. Skeletal muscle attaches to the bone by tendons, and together they produce all the movements of the body. The skeletal muscle fibers are crossed with a regular pattern of fine red and white lines, giving the muscle a distinctive striated appearance. Hence they are also known as striated muscle.

Structure and Function

The skeletal muscle is one of the three significant muscle tissues in the human body. Each skeletal muscle consists of thousands of muscle fibers wrapped together by connective tissue sheaths. The individual bundles of muscle fibers in a skeletal muscle are known as fasciculi. The outermost connective tissue sheath surrounding the entire muscle is known as epimysium. The connective tissue sheath covering each fasciculus is known as perimysium, and the innermost sheath surrounding individual muscle fiber is known as endomysium. Each muscle fiber is comprised of a number of myofibrils containing a number of myofilaments. When bundled together, all the myofibrils get arranged in a unique striated pattern forming sarcomeres which are the fundamental contractile unit of a skeletal muscle. The two most significant myofilaments are actin and myosin filaments which are arranged distinctively to form various bands on the skeletal muscle. The stem cells which differentiate into mature muscle fibers are known as satellite cells which can be found between the basement membrane and the sarcolemma (the cell membrane surrounding the striated muscle fiber cell). When stimulated by growth factors, they differentiate and multiply to form new muscle fiber cells.

The primary functions of the skeletal muscle take place via its intrinsic excitation-contraction coupling process. As the muscle is attached to the bone tendons, the contraction of the muscle leads to movement of that bone that allows for the performance of specific movements. The skeletal muscle also provides structural support and helps in maintaining the posture of the body. The skeletal muscle also acts as a storage source for amino acids that can be used by different organs of the body for synthesizing the organ-specific proteins. The skeletal muscle also plays a central role in maintaining thermostasis and acts as an energy source during starvation.

Embryology

Distinct transcriptional mechanisms and specific gene regulatory activity control the differentiation of muscle fibers. During embryogenesis, it is the para-axial mesoderm that undergoes stepwise differentiation to generate the muscle tissue. The para-axial mesoderm on either side of the neural tube starts to differentiate and undergoes segmentation to form the somites. The somites get stimulated by the myogenic regulatory factors to differentiate into a dermomyotome and sclerotome. These regulatory factors include the Wnt, Shh and BMP4 proteins. The neural tube and the surface ectoderm are the primary sources of Wnt proteins, the Shh proteins (Sonic Hedge Hog) source from Notochord, and the lateral mesoderm plate produces the BMP4 protein. The lateral aspect of the dermomyotome undergoes epithelial to mesenchymal transition as it proceeds to migrate on the ventral side to form a unique myotome below the dermatome.

The myotome then differentiates to form the skeletal muscles in the body after receiving stimulation from the Sonic Hedgehog (Shh) signaling molecule from the notochord that results in the Myf5 expression and subsequent differentiation. The dorsomedial aspect of the myotome differentiates into epaxial myotome giving rise to back muscles. The ventrolateral aspect differentiates into hypaxial myotome that gives rise to muscles of the body wall.

Several signaling molecules like the Wnt and the BMP along with some transcription factors like Sine Oculis homeobox are responsible for this differentiation. The development of skeletal muscles in the limb and the trunk depends on the expression of MyoD and Myf5 and their effects on the different myoblasts. These embryonic myoblasts undergo further differentiation to form the primary muscle fibers and eventually secondary myofibers by the union of myoblasts in the fetus. After birth, the satellite cells act as stem cells and are responsible for the further growth and development of skeletal muscles.

Blood Supply and Lymphatics

The main artery or the primary artery supplying blood to the skeletal muscle courses parallel to the longitudinal axis of the muscle fiber. The primary artery gives off tributaries which are known as feed arteries, that are perpendicular to the primary artery and proceed towards the external connective tissue sheath of the muscle fiber called perimysium. The feed artery branches into primary arterioles which after two more orders of branching gives rise to transverse arterioles, which in turn give rise to terminal arterioles. The terminal arterioles are the final vascular branches, and they perfuse the capillaries that are present within the endomysium and travel parallel to the longitudinal axis of the muscle fiber. The terminal arteriole along with the capillaries that it supplies is known as a microvascular unit, and it is the smallest unit in the entire skeletal muscle where the blood flow can be regulated.

Lymph capillaries originate in skeletal muscle in the microvascular unit within the endomysium near the main capillary bed and drain the tissue fluid. These capillaries merge with each other to form the lymphatic vessels as they drain out the tissue fluid. These lymphatic vessels go through the perimysium and join with the larger lymphatic vessels. Unlike the blood vessels, the wall of the lymph vessels within the muscle does not have contractile property due to lack of smooth muscles (in the wall), so they depend on the muscle movement and arteriolar pulsations to drain the lymph out.

Nerves

The neuronal innervation of a skeletal muscle typically comprises of sensory nerve fibers, motor nerve fibers, and the neuromuscular junction. The nerve fibers are composed of myelinated as well as non-myelinated nerve fibers. The cell bodies of the neurons give rise to large axons which are generally unbranched and travel to the target muscles for innervation. Near the target muscle, the axons divide into multiple smaller branches to innervate multiple muscle fibers. The motor nerve terminal has abundant mitochondria, endoplasmic reticulum and numerous membrane-bound synaptic vesicles containing neurotransmitter- acetylcholine. Once the action potential travels to the neuromuscular junction, there occurs a series of processes culminating into the fusion of the membrane of the synaptic vesicles with the presynaptic membrane and subsequent release of the neurotransmitter into the synaptic cleft.

The postsynaptic membrane of the muscle fibers has a massive concentration of neurotransmitters (AchR) receptors. These receptors are transmembrane ligand-gated ion channels. Once the neurotransmitter activates these ion channels, there is a rapid depolarization of the motor end plate, which initiates an action potential in the muscle fiber resulting into muscle contraction.

Biophysics of the Generation of EMG Signals

Generation of Action Potential

In experiments involving insertion of a micro pipette electrode inside the muscle fiber it's found that the cell is negative inside with respect to the extracellular environment outside. The resting membrane potential of 70 to 90 mV is mainly maintained by the presence of the sodium potassium pump (Na^+ - K^+ pump), that continuously works against the concentration gradient of ions flowing through the membrane. With the arrival of action potential at the presynaptic membrane, acetyl choline is released that passes through the synaptic cleft and excites the post synaptic muscle fiber membrane, producing a local potential gradient. The excitation is called depolarization. It's during this phase when there is rapid inflow of positive ions (Na^+) that renders the inside of the cell positive. The outside of the membrane is negatively charged so positively charged ions from the neighboring areas, will get attracted towards it. As depicted in figure below:

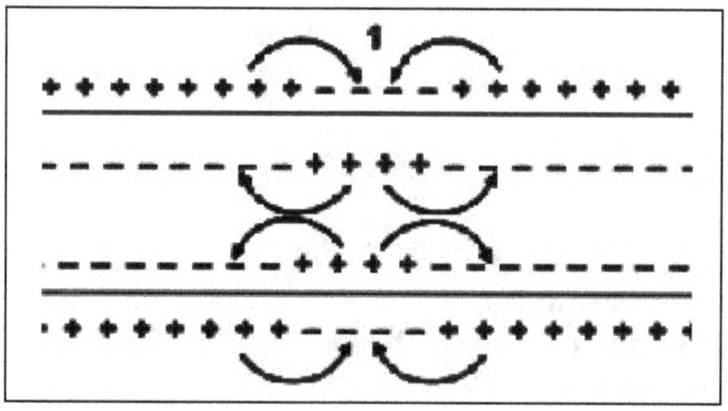

This process continues and the propagation of action potential in this manner is called electrotonic spread. Theoretically speaking this spread can occur in any direction, this holds good in the case of muscle, but in nerves the electrotonic propagation is unidirectional, because of unidirectional transmission in neuromuscular junctions and synapses. Every wave of depolarization is followed by repolarisation. And this process repeats itself over and over again till it reaches the tendon endings. We know that a current flow is also associated with potential changes. This propagating action potential creates a transmembrane current profile that also travels along the membrane of the muscle or sarcolemma. For muscles fibers, the length of depolarization and repolarization are very small, i.e. in millimeters. So we can consider a muscle fiber a very thin tube, in which current flows only axially, in other words it can be considered as a line source model. Figure below is the line diagram of the linear core conductor model.

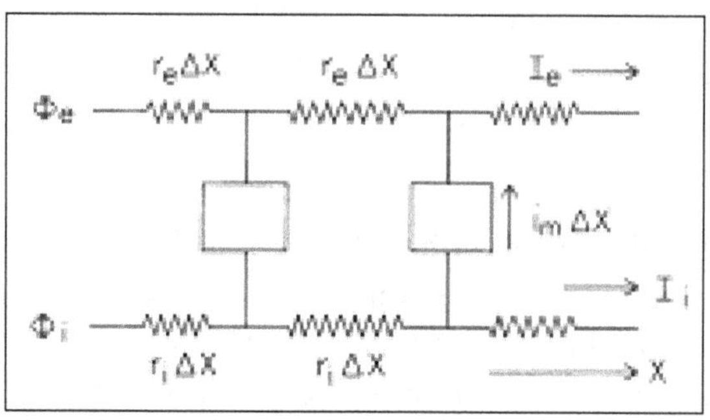

Here Δz is a assumed to the center of the fiber, the subscript i represents intracellular and subscript e represents extracellular, r represents resistence, I represents current and Φe represents extracellular potential and $\varnothing i$ represents intracellular potential. Now the decrease in potential per unit time is equal to product of resistance per unit length and current passing through it. For intracellular it's given as:

$$\frac{\delta \varnothing i}{\delta z} = -I_i r_i$$

For extracellular it's given as,

$$\frac{\delta \varnothing e}{\delta z} = -I_e r_e$$

The motor unit consists of the motor neurons and the muscle fibers. When an action potential is generated in the neuron, it activates the muscle fibers associated with it. The degree of delporization of the different fibers varies based on the distance of each fiber, from the source. The summation of all the potentials is called as Motor Unit Potential (MUP). The electrodes record this potential and depict it as the EMG on the computer interface.

Concept of Volume Conductor

The biological tissues separating the sources and recording electrodes are called the volume conductor. It plays a vital role in the acquisition of signals. The generated action potential creates an electric field in the surrounding space, thus the potential generated by the motor unit can also be detected at a distant location. The potential signals have to traverse through various media so as to reach the recording electrodes. Under static conditions the volume conductor obeys some relations,

$$\Delta . j = I$$

$$J = \sigma E$$

$$E = -\Delta \varnothing$$

Where J is the current density in the volume conductor, I is the current density in source, σ represents the conductivity of the media, if the media is homogenous, conductivity doesn't depend on the site of recording. But if non homogenous, then the conductivities across various media and distance of electrode and source play a very crucial role.

Concept of Cross Talk

Cross talk is the signal that is detected over a particular muscle, but it is actually generated by a nearby muscle. It is the far field signals that are being generated by the dying of the potentials at the tendons. This phenomenon is mainly seen in surface recordings. The properties of volume conduction comes into play when the "far field" signals travel along the surface and tissues. The main problem with the phenomenon of crosstalk is that they influence the interpretation of the EMG recordings. Some important points of crosstalk that we must remember:

- It's mainly by the far field signals which are the results of the extension of the potentials at the tendons.

- When we talk about the shape of the signals over the active muscle, the crosstalk signals are very different from the signals of the active muscles.
- The bandwidth signals of the crosstalk are larger than the signals from the active muscle.

The Components of EMG Instrument

To measure the emg of the muscle, the recording apparatus requires: 1) Electrodes 2) Amplifier 3) Recording System.

Electrodes

The various electrodes involved in the measurement are: 1) The surface electrodes 2) Insertion electrodes. 3) More advanced are the 1 and 2 Dimensional electrode arrays.

- Surface electrodes: They are applied superficially. They can be either disc or cup electrode. They are very useful in superficial recording of the muscles. But they also pick up global recordings. Which makes them less precise in recordings.
- Insertion electrodes: They are either needle type or wire type. They are invasive in nature but are very precise. Being precise they find their use in research.
- Electrode arrays: They work as multichannel EMG. Here many electrodes are arranged into an array and they pick up signals from multiple points and they give spatially sampled image or map of the instantaneous monopolar surface potentials. The extractions of information from these signals give very accurate interpretation of the muscular activity.

The proper placement of electrodes reduces electrode skin impedance- The electrodes that we using these days come with already attached gel. When the electrode is applied on the skin, many charge layers are created (also called as Helmholtz layer) at the metal gel and gel skin interface, thereby forming a imagenary (or capacitive) component of impedance. If we record the impedance between two electrodes separated by gel and the between the gelled electrode applied on the skin, we will find that the impedence in the first scenario is lesser than the second. To reduce the impedence, it's also important to prepare the skin surface.

Impedence and effects of skin preparation: The electrode - gel- skin impedence affects the signal acquision. The widely varied impedence interferes with the monopolar or bipolar recordings. So having a good skin preparation is desirable.

The best mode of the skin preparation is:

- Rubbing the surface with an abrasive conductive paste and rinsing.
- Washing the surface with soap for 30seconds and then rinsing.

These decrease the impedence and improve the signal acquisition.

- The Ampifier: A high gain amplifier is used which has a selectivity for frequencies from 10 to 5000 Hz.

- The recording system: In any design of a EMG signal acquisition system, the three principal requirements must be kept in mind.
- The front end: It interfaces the analog input signal to the analog to digital converter (ADC).
- The voltage reference: It is an electronic device which maintains a constant voltage. The constant voltage is maintained regardless of the temperature changes, the load on the device, power supply variations and passage of time.
- Digital interface: It's the medium through which we interact with the computer. The interaction act be in form of auditory, visual and functional components (touch or talk).

The types of recordings in electromyography that can be found in a subject:

- The insertional recording: They are mainly seen while inserting the wire or needle electrode. They should not be considered in the analysis. The presences of these recordings are evidence of the procedure.
- Resting muscle recordings: These are the types of recording while the muscles are at rest. It is believed to be the evidence for activity of the neuromuscular activity. In recent times it's considered to the basis of muscle tone.
- The muscular activity recordings: The recordings that is associated with the voluntary contraction of the muscles. The EMG signals signals seen while muscle contraction is the motor unit action potential (MUAP) or in simple words it's knows as motor unit potential (MUP).

The factors affecting the signals recordings:

- Movement of the cables.
- Momentary loss of skin electrode contact causing motion artifact.
- Fluctuations of the charge distributions on the skin and connection cables.
- Power line interference.

Different Types of Recordings System in EMG

Monophasic Recording

It's mainly done with monopolar electrode. In mono polar electrode, the active electrode is placed on the muscle and another electrode is maintained at a constant potential, but is located far from the active electrode (indifferent electrode).

A monophasic electrode mainly records the absolute potential because the indifferent electrode is mainly maintained at constant potential. To begin with:

- The muscle fiber is polarised and the wave of depolarization has not reached the segment of muscle. So, there is no reading.
- The wave of depolarization now arrives at the muscle segment. The polarity changes and excursion is picked up by the electrode.

- The wave of depolarization passes away from the muscle and the membrane is repolarized now. The excursion recorded earlier, now moves towards the baseline.

- The action potential has moved away from the vicinity. The recording now reaches the baseline and that's how a monophasic recording takes place.

Biphasic Recording

In EMG we also use Bipolar electrodes. The recordings show the potential difference between the tips of the electrodes. Here two active electrodes are placed side by side at very close vicinity. The recording is biphasic.

- Let's assume the shaded area is the nerve impulse. In the initial phase the muscle fiber is polarised. The activity that we get is considered the baseline. As both the electordes are at the same potential.

- The arrival of action potential (wave of depolarization) reverses the polarity. The 1st electrode becomes negative in respect to the 2nd electrode. In EMG the electrodes are placed extracellularly, so the first electrode will give an excursion in the recording.

- The wave of depolarization now reaches the 2nd electrode. Here both the electrodes are in the same polarity. So the recording comes to the baseline.

- Now 1st electrode is undergoing repolarization, which makes it positive in respect to 2nd electrode (which is still in depolarized state). This will lead to an excursion in the opposite polarity.

- The 2nd electrode is repolarized now, so both the electrodes are at the same polarity. The recording comes to the baseline.

- Thus a single action potential gives rise to a biphasic recording.

Ultrasonography

Ultrasonography (USG) is a commonly used imaging technique used in medicine for diagnosis of different disorders. It may also be used as a therapeutic modality in physical therapy and rehabilitation as it produces heating effect and may relieve pain and swelling.

Sound Wave Physics

Sound wave is essentially a pressure wave that consists of alternate compressions and rarefactions. Production of sound waves requires a source. When a sound wave is produced, particles of the medium start vibrating and relaxing. One important property of sound is that it requires a medium to travel, which may be solid, liquid or gas. So, sound waves cannot travel in vacuum.

Human ear can perceive wavelengths in the range of 20Hz to 20,000 Hz (1 Hertz (Hz) is 1 cycle per second). This frequency range is known as the auditory frequency range. Some animals such as bats can hear sounds well beyond the upper limit of human audible frequency. They use the same for navigation, which we will discuss further in this module when we talk about Doppler principle.

One question that should strike us now is why do we call Ultrasound waves as 'Ultra'-sound waves? In common parlance, ultrasound means sound waves at very high frequencies (hence the use of the term 'ultra'). In common practice, ultrasound waves are more than 20,000 Hz in frequency.

Basic Physics Behind Ultrasound

Ultrasound waves are also governed by physical laws that govern sound waves. US waves require a source for production. Also, as with sound waves, Ultrasound waves also require medium for travelling. The medium may be solid, liquid or gas. When striking a medium, some waves are reflected, some transmitted while others are refracted. Higher frequency of US permits better spatial resolution but reduces the depth of penetration.

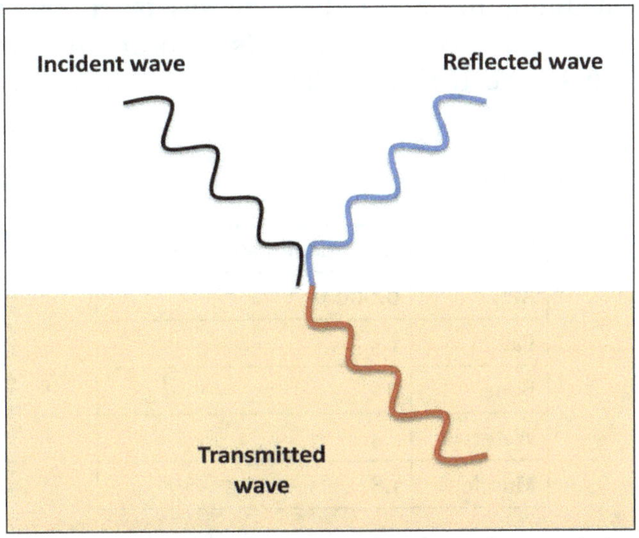

US waves when striking a medium exhibit reflection and transmission. This may be decided by the property of the interface between the media.

How are US Waves Produced

The working of ultrasound is based on piezoelectric effect, which was discovered by Curie brothers in 1880. First let us understand the meaning of piezoelectricity - the term is derived from the Greek work piezein, which means to apply pressure. The effect is seen in quartz crystals that change shape when electricity is applied to them. This change of shape is very rapid and is associated with production of high frequency sound waves. The reverse phenomenon is also observed - the crystals also develop electric field on application of pressure. This allows the crystals both to emit sound waves as well to respond to them. This piezoelectric effect is at the heart of USG and is utilized in US probes to produce high frequency sound waves.

Important Terminologies Related to USG

Power and Intensity

Power is the amount of acoustic energy per unit time in US wave. It is expressed in the units Watts.

Power per unit area is known as intensity. It is expressed as Watts/cm2. Since US beam has energy associated with it, the same is transferred to the tissues during USG. Therefore US beams are delivered in short bursts known as pulses during USG and this modality is known as Pulsed USG.

Acoustic Impedance

There are some terms that should be understood in relationship to ultrasonography. One of the most important of them is Acoustic impedance. In physical terms, acoustic impedance is the product of velocity of sound and the density of the medium. To make things simpler, let us break the term into two components – acoustic and impedance. Acoustic refers to sound waves and impedance is the measure of resistance encountered. So, it is a measure of the resistance encountered by the ultrasound waves when passing through a medium.

The resistance encountered is proportional to the density of the medium – therefore when US waves penetrate a dense medium, they encounter high resistance. When traveling through medium that has different acoustic impedances, there is reflection of some waves and refraction of other waves. Typical values of acoustic impedance of some materials are mentioned in Table below.

Table: Acoustic Impedance of common media.

Medium	Acoustic Impedance ($x10^6$)
Air	0.0004
Fat	1.5
Bone	8
Water	1.4
Muscle	1.8

Acoustic impedance of bones and enamel is amongst the highest in the body. Therefore whenever US waves strike them, most of them are reflected and almost none pass through. This is the reason for poor visualization of hard tissues like bones and enamel in USG.

Attenuation Constant

Attenuation constant is another important term related to USG. Attenuation means loss of intensity. When travelling through a medium, the reduction in intensity of US amplitude is proportional to the distance travelled in the medium. This knowledge of attenuation of US is important to us as it can help us to adjust the amplitude of the US waves to compensate for loss of signal when it travels through soft tissues.

Doppler Effect

Doppler effect is employed in USG to enquire about blood flow. This is based on the Doppler principle which states that there is relative change in the frequency of the sound waves heard by the observer on the basis of the relative movement between sound source and observer. A simple explanation of the same is given in figure.

Schematic representation of Doppler effect.

US probes utilize this effect to detect blood flow. We know that flowing blood is essentially a flow of cells suspended in plasma. These cells when struck by an US beam lead to change in frequency of reflected waves based on the velocity of flow. This change in frequency is detected by the probe and converted into velocity and flow based on some mathematical calculations.

How to Perform Ultrasound

Performance of USG is relatively simple. As with any medical procedure, informed consent is obtained from the patient/subject before starting the procedure. When performing USG for a female patient, we should always ensure presence of female attendant before starting the procedure.

Part of the body to be examined is exposed adequately. It is followed by application of a coupling agent on the area. A coupling agent (also known as conducting jelly) is a water based gel that ensures optimum transmission of the US waves into the area of interest. The power may be adjusted accordingly to visualize the organ/structure of interest. We always go by the ALARA principle – as low as reasonably allowable – when adjusting power for performing USG. Representative image of the equipment, gel and procedure is depicted below:

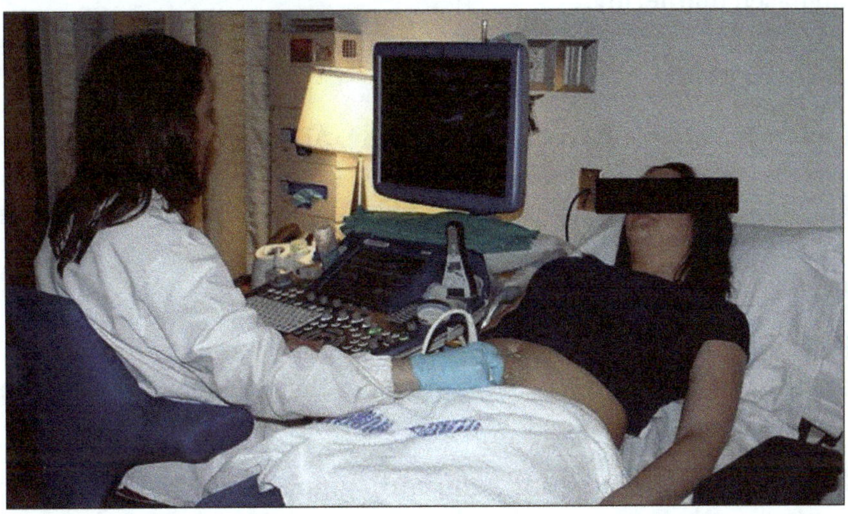

Applications of USG

Applications of USG can be broadly categorized into diagnostic (for the purpose of knowing the disease affecting the patient) and therapeutic (as a treatment modality). The same are summarized in figure.

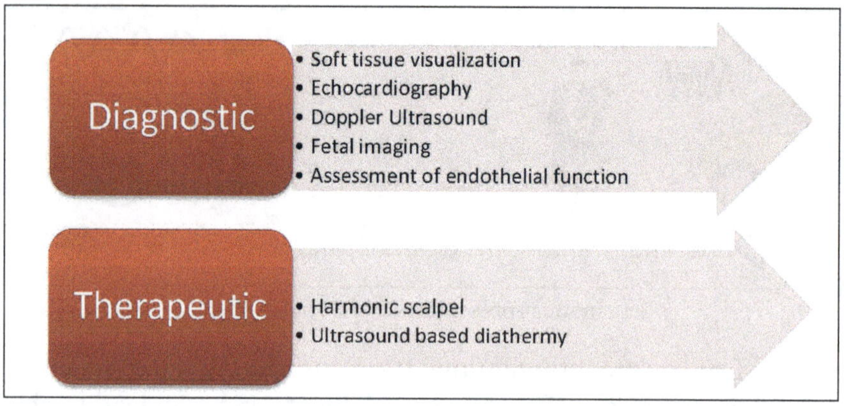

Diagnostic

- Soft tissue visualisation: Abdominal, transvaginal, transrectal, ocular.
- Echocardiography.
- Doppler ultrasound: Assessment of cerebrovascular reactivity - Transcranial Doppler.
- Fetal imaging.
- Assessment of endothelial function: Flow mediated dilatation.

Therapeutic

- Harmonic scalpel.
- Ultrasound based diathermy.

Soft Tissue Visualization

US is routinely used to visualize soft tissues of the body. Abdominal viscera like liver, gall bladder, spleen, kidney, ureters, urinary bladder are visualized in real time. We can assess the size of the organs, contents of hollow viscera (like stones and fluid) and their movement in real time.

Echocardiography

Echocardiography is visualization of different chambers of the heart and valves inside. We can observe the contractile function of the heart, the size of the chambers, the movement of blood in different phases of cardiac cycle and the motion of the heart valves.

Two types of USG techniques are used for Echocardiography – 2D mode and M mode. 2D mode visualizes the heart in two dimensions while M mode, or motion mode, helps detect the motion of continuously moving structures such as wall and valves. Different acoustic windows are available for Echocardiography.

- Suprasternal (beam projected from above sternum, directed downwards).
- Apical (beam directed from apex of the heart).

- Parasternal (beam directed from area adjacent to the sternum).
- Subcostal (beam directed upwards from below sternum).

These windows are unique in helping assess the heart from different planes and are helpful in identifying different structures. Echocardiography is very useful is assessing the following:

- Contractile function of the heart.
- Lesions of the valves–stenosis (narrowing) and regurgitation (leak).
- Dimensions of chambers.

Doppler Ultrasound

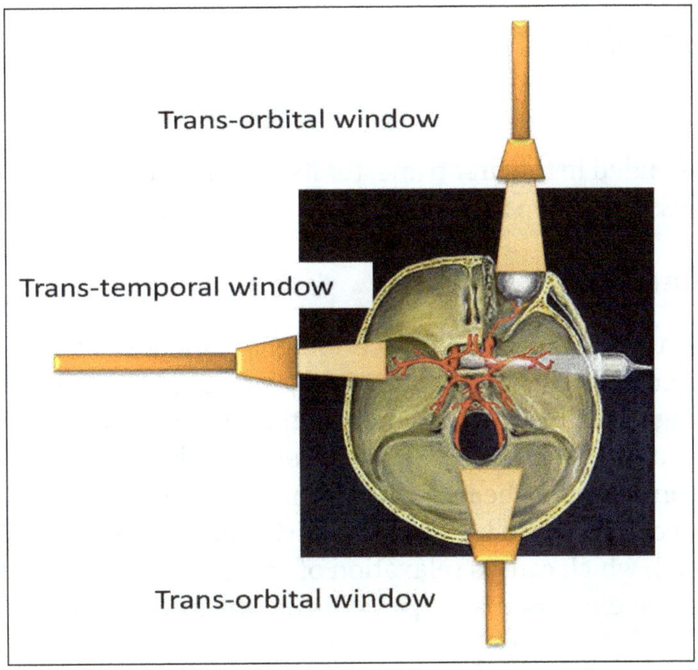

Transcranial Doppler with bony windows for probe placement.

Doppler USG employs Doppler principle to detect frequency shift based on relative motion between sound source and target. There are different modalities of Doppler USG:

- Continuous wave Doppler: A pair of crystals is used for USG. One crystal emits sound waves while other detects them. The emission and detection process is continuous and therefore very useful if the frequency shift is very fast.
- Pulsed wave Doppler: The sound beam is emitted in small pulses rather than continuously. The technique is commonly employed to reduce the amount of energy delivered to the tissues and therefore leading to low heat production.
- Color flow mode: This technique employs a color scale to depict the direction of flow of blood towards or away from the probe. Red depicts flow towards the probe, while blue depicts flow away from the probe.

At our department, we use Transcranial Doppler as a tool for visualization of the cerebral circulation – the circle of Willis. We insonate the cerebral arteries through the thinner region of skull (known as acoustic windows) and calculate blood flow through them. The variation of blood flow can be assessed in response to physiological parameters such as hypoxia and blood pressure fluctuations.

Fetal Ultrasound

Fetal imaging is routinely performed as a part of Antenatal checkup in pregnant females. It is helpful in assessment of the following:

- Number of fetuses: Singleton pregnancy or multiple fetuses.
- Fetal viability.
- Establishment of gestational age.
- Fetal defects.

The first scan is recommended in the first trimester itself. The combination of B mode and M mode is recommended to assess the above mentioned growth parameters.

Flow Mediated Dilatation

Flow mediated dilatation (or FMD) is a test of endothelial function. In the FMD protocol followed at our department, we occlude the brachial artery by using a blood pressure cuff. The pressure employed is 200 mm Hg (which is much higher than systolic blood pressure) and the occlusion is maintained for 5 minutes. When the occlusion is released, blood starts flowing rapidly into the forearm and leads to considerable shear stress on the vessel wall. This leads to dilatation of the vessel due to production of a chemical known as Nitric Oxide from the endothelium (the innermost layer of the blood vessels), which causes relaxation of smooth vessels of the blood vessel wall. This change in diameter can be expressed as a percentage which helps us assess the function of the endothelium.

Advantages of Ultrasound

- No risk of radiation. X rays, CT, PET.
- Bedside tool.
- Patient acceptability.
- Low cost.
- Multiple applications.

Ultrasound is one of the most commonly used imaging modalities today. There are multiple advantages of using USG. First and foremost is that it does not expose the patient to ionozing radiations. USG consists of sound waves that are harmless and do not pose any major risk to the patient. We can contrast the same with X Rays and Computerised Tomographic Scan (CT) tat expose the subject to ionising radiations and therefore are not advisable at short intervals. Multiple exposure to ionising radiations carries the risk of malignancy in later life as thay may create mutations in the subjects.

Another advantage of USG is portability. Ultrasound machines available today are of slightly larger than commerically available laptops and can be mounted on a trolley and be shifted anywhere as and when the need arises. Therefore USG can be done in emergency room and bedside conveniently without shifting patient to specialised testing rooms.

Patient acceptability is another reason for popularity of USG. Since the technique requires only placement of a probe on the skin surface with application of a jelly, it is quite acceptable to the majority of patients. The cost of a USG is quite reasonable with respect to other sophisticated imaging techniques such as CT and MRI scan.

Applications of USG are enormous. Based on the mode discussed previously, USG can be used to visualise soft tissues, organs, blood vessels and even foetus inside the uterus of an expectant mother. Therefore USG finds applications in different scenarios in health and disease.

Disadvantages of Ultrasound

- Operator dependent: Performance and interpretation depends on examiner's skills.
- Poor for bony tissues or air filled cavities. Useful for soft tissue visualisation only.

Operator dependence is one of the major drawbacks of USG. By this term, we mean that performance and interpretation of USG is dependant on the skill of the person performing the procedure. It takes long hours of training and practice before one becomes an expert in performing the USG procedure as well interpreting the records obtained. In some cases, parameters obtained on USG may be subject to observer bias - the measurement of parameters depends on the person performing the measurement. This is gradually being overcome by software assisted image analysis that obviates observer bias.

The reflection of ultrasound waves depends upon the presence of interfaces. If the USG beam travels from solid to solid or from solid to gas, there is considerable reflection of waves and the signal obtained is quite poor. This situation can be encountered when visualizing dense tissues like bone or air filled cavities like bowel loops.

Blood Flow Measurement

Blood volume accounts for 7% of human body weight with an average density of approximately 1.054g/cm^3 and is a specialized fluid consisting of red blood cells, white blood cells and platelets suspended in a fluid medium known as blood plasma. An average adult has a blood volume of roughly 5 liters. The blood conveys oxygen from the lungs to the tissues, carbon-dioxide from tissues to lungs, supplies nutrients to the tissues, removes metabolic end products and performs immunological functions. Additionally, blood regulates pH and body temperature and maintains the water content of the tissues. Diseases/disorders of the blood that commonly affect human beings can involve any or all of the constituent blood cells. Other types of blood diseases affect the proteins and chemicals in the plasma that are responsible for clotting.

One of the best indicators of a patient's health is the concentration of oxygen and other nutrients in cells which is the primary measurement any physician would like to acquire from a patient and closely relates to the amount of blood flow. Research on the various measuring techniques of blood flow has a direct impact on our improved understanding and management of human health. Measurement of blood flow and changes in blood volume are of vital importance and provide essential information for the diagnosis of diseases with impaired blood flow.

The measurement of blood flow and its related parameters have been evolved through a number of techniques involving different principles. From the start of 19th century, till now, a number of attempts have been made to modify the existing methods as well as establishing novel concepts in this field.

Early Methods for Blood Flow Measurement

The history of blood flow measurement started from the mid of 19th century when the ends of a known volume air tight U-tube was connected to a blood vessel to measure the rate of arterial blood flow according to the Poiselle formula. Although these methods were applied for estimating the volume rates of blood flow in the limbs, kidneys, head and abdominal viscera of average size laboratory animals, cannulation of the blood vessel and the resistance offered by these methods to the actual blood flow were considered to be the disadvantages of such measurements. Collection of venus outflow in a graduated cylinder over a known time interval was also another method used in early times to estimate the flow, which was subjected to automation in its successive modifications. Introduction of drop recorders helped measurement of flow rates lower than 2ml/min. The disadvantages of drop recorders such as blood loss and reaction at the electrodes were overcome by the design of closed circulatory system and use of photoelectric method for the measurement of flow. Effects caused by air in the measuring chamber in the rate of drip were also eliminated by replacing air with silicone fluid, thus providing a regular drip rate.

In yet another attempt, the time taken by a deliberately introduced bubble to move a known distance along a glass tube having similar diameter as the bubble through which blood is allowed to flow was estimated to provide a measure of the flow velocity. Bubble flow-meters were known for their simplicity, reliability, low resistance to flow, and mostly linear calibration yet with the disadvantages such as cannulation, anticoagulants and the possibility of slip between the bubble and blood.

Contemporary to the cannulated measurement of blood flow using glass tubes, trails were also made for measuring blood flow using dye dilution techniques. The principle involved in this method is available elsewhere. Experiments were carried out using continuous/bolus injection of the dye to the blood stream. Factors affecting the end result using this method include errors in blood sampling procedure, nature of dye infusion and type of dye being used. Typical applications include limb blood flow, portal blood flow renal blood flow, retinal blood flow etc. Injection of radioactive microspheres into the systemic circulation and measuring the concentration variation was also practiced as an indicative of the blood flow. The use of radioactive elements in the blood stream was surely invasive upon repeated testing and subsequently this method has only been utilized to estimate flow in animals.

Measuring flow based on the electric conductivity of blood was another alternative to be tried out in the past years. Related theories were developed in early attempts, making use of permanent/ DC magnets and suffered from disadvantages such as interfering potentials and electrode polarization. AC magnets were used as an alternative, with different excitation waveform shapes such as sinusoidal, square, trapezoidal etc. with respective pros and cons yet with the disadvantage of "transformer effect" at the electrodes. Alternatively, thermal convection principle was applied to the blood flow studies, which made use of thermal sensors like thermocouples and thermistors. The working principle is that when a hot object is placed in a colder flowing medium it gets cooled due to thermal convection and the rate of cooling is proportional to the rate of flow of the medium. The electrical energy required to maintain the constant temperature of a temperature sensor introduced into the stream is taken as the measure of the blood flow rate which is found out by the help of equations. Most of the times, these sensors will be a part of a bridge circuit such as in hot film anemometer. The concept of quantifying the oxygen content in blood and using it for the measurement of blood flow began with the development of an oxyhemograph, which worked on the principle which was used to measure the circulation time (CT) as fast as few tenth of a second in animals such as albino rats and dogs irrespective of their skin colour from which the blood flow velocity was estimated.

Volumetric changes in the blood flow has been achieved earlier times by direct Plethysmography techniques that involved trapping the body part of interest in a fixed pressure chamber and monitoring the volume changes of the chamber occurring due to the changes in blood flow. Different types of transducers were made use, to serve this purpose and the designed plethysmographs were subsequently known by the type of transducer involved. Capacitance plethysmograph which made use of the change in capacitance of the transducer due to the volumetric changes in flow and the impedence plethysmograph which measures small changes in electrical resistance of the region of the body part of interest are only few examples of these types. The down side of this is the slow response time and the accuracy.

Recent Advancements

Disadvantages of conventional methods led to the research and development of new techniques in the measurement of blood flow.

Ultrasound Technique

Ultrasound waves (frequency range above 20kHz) have found tremendous application in different fields of medicine such as imaging, localizing tissues as well as in flow measurements. This technique is used for blood flow measurement based on the principle of propagation of ultrasound waves in the tissue. Generally, these ultrasound systems are categorized as conventional diagnostic ultrasonic imaging systems as well as Doppler ultrasound systems. In conventional imaging, the typically used frequency is in the range of 2MHz to 15MHz.

Doppler Ultrasound

Doppler ultrasound is a technique used for making non-invasive measurements of blood flow. The underlying principle is that a frequency shift occurs when sound is reflected from a moving source. To measure the velocity of blood an ultrasound of a specific frequency is transmitted into a blood vessel as shown in figure below and the sound that is reflected back from the blood vessel is detected. Due to the movement of blood the sound undergoes a frequency (Doppler) shift that is described by the equation below where, c=1540m/s is the velocity of sound in the test tissue,

$$v_D = \frac{cf_D}{2f_o \text{Cos}\theta}$$

v_D is the blood flow velocity, f_o is the irradiated ultrasound frequency, f_D is the Doppler shift frequency of the received sound signal and θ is the angle of placement of the transducer with respect to the blood vessel. The transmission of sound signal can be either in the continuous wave (CW)/ pulsed wave (PW) mode.

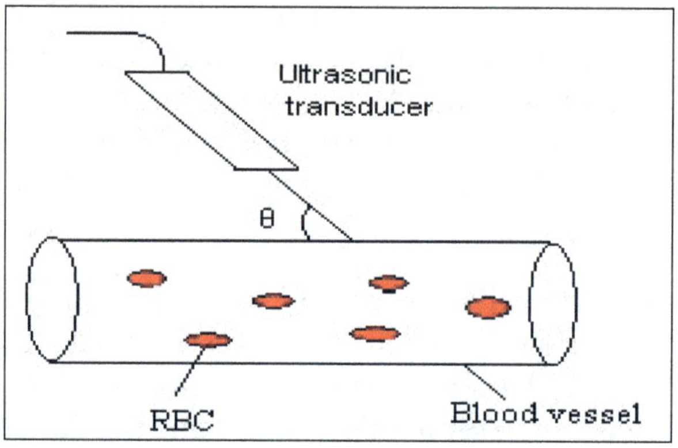

Doppler ultrasound method.

Different Types of Doppler Ultrasound Flowmeters

Continuous Wave Doppler Ultrasonic Flowmeters

As the name suggests, continuous wave Doppler ultrasound systems use continuous transmission and reception of ultrasound waves. Doppler signals are obtained from all vessels in the path of the ultrasound beam until the ultrasound beam becomes sufficiently attenuated during its propagation along the tissue.

Several studies were carried out based on the principle of Doppler ultrasound using continuous waves. Changes in blood flow velocity in the carotid artery during different stages of physical exercise was studied based on the principle of ultrasound using a 2MHz CW Doppler ultrasonic probe and a Doppler discriminator circuit. The schematic of the measurement system is shown in figure below. The protocol exercise using a cycle ergometer with transient response and ramp response were conducted. In the Doppler signal discriminator, the Doppler signal and the 2MHz fundamental frequency are detected synchronously. The low-frequency and harmonics noise were then removed so that the Doppler signal is taken out from its range and digitized using an analog/digital converter interfaced to the computer. The blood flow velocity was then calculated using equation above. In the transient response experiment the maximum blood flow velocity increased initially as the exercise started and then decreased when a constant workload was applied and increased again immediately before the end of the physical exercise. In addition, an arrhythmia was also detected to occur in the subject immediately after exercise. In the ramp response experiment, the blood flow velocity precisely increased during exercise.

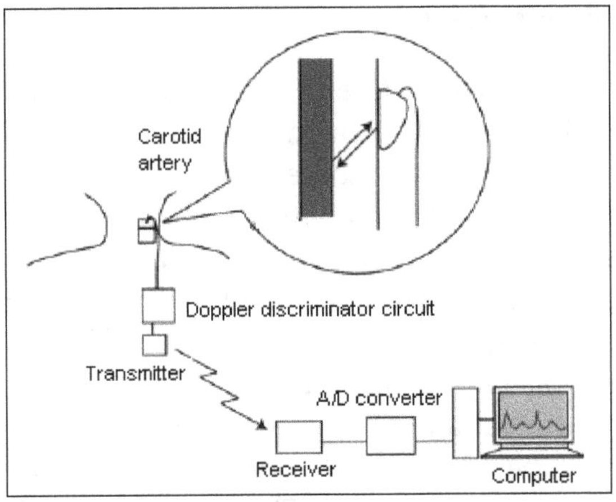

Ultrasound based blood flow measurement system.

Following the measurement in a single artery, the variations of the maximum blood flow velocity (MBFV) were evaluated simultaneously in the carotid, brachial and femoral arteries during postural changes. In the experimental setup as shown in figure below, three different probes collected data from the arteries and variations in MBFV during passive postural changes in supine position, 900 head-down-tilt (HDT) and 900 head-up-tilt (HUT) were measured. MBFV in brachial and femoral arteries were found to increase during supine to HDT posture and were found to decrease during supine to HUT posture, whereas very small changes in MBFV were observed in the carotid artery. The increase and decrease in MBFV are expected to be an effect of venous pressure and a constant MBFV in carotid artery was attributed due to cerebral auto-regulation.

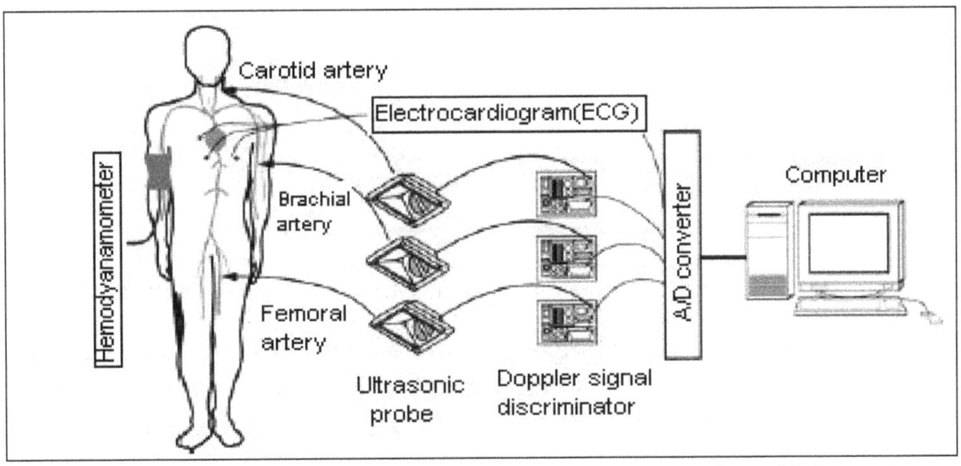

Ultrasound velocimeter system for simultaneous measurement in multiple arteries.

Analysis of the dynamics and distribution of the blood circulation in human subjects during exercise was also studied. By changing exercise postures to supine, seating, standing as well as walking on treadmill, and by analyzing the obtained power spectrums, flow velocity in the carotid, brachial and femoral arteries were measured. Lower MBFV values and higher wave propagation time were observed in carotid artery during standing position as compared to supine and sitting position. The carotid artery showed higher flow velocity values in proportion to the work rate during walking on the treadmill. The blood flow measurement during physical exercise was found to give useful data in the early detection of cardiovascular diseases which could not be detected by ECG due to the measurement of changing flow in the carotid artery directly.

Ultrasound for Measuring Perfusion

In general, ultrasound techniques are effective in detecting flow velocities in major blood vessels with considerable blood velocity. Capillary flow measurement which is generally known as perfusion or tissue blood flow measurement becomes difficult with ultrasound as the flow velocity reduces, the discrimination/analysis of the reflected spectrum from the noise spectrum becomes difficult. In one of the few efforts, a CW Doppler ultrasound equipment operating at a slightly high frequency of 8.1MHz was used to detect low blood flow velocity in the microcirculation in the thumb of a subject. The measurements were taken using the ultrasound probe with and without the target (to account for the noise spectrum) and the results were compared with the readings taken with a disk phantom rotating at a low velocity. The experimental schematic is shown in figure below. In case of the free probe, the spectrum showed noise spikes at low frequencies (below 20Hz) due to the tremor in the operator's hand. The amplitude of the spectrum from the thumb was higher than that of the free probe in the 20-100Hz region which suggested that additional Doppler shifted energy was being reflected back to the probe, indicating the statistics of the microcirculation.

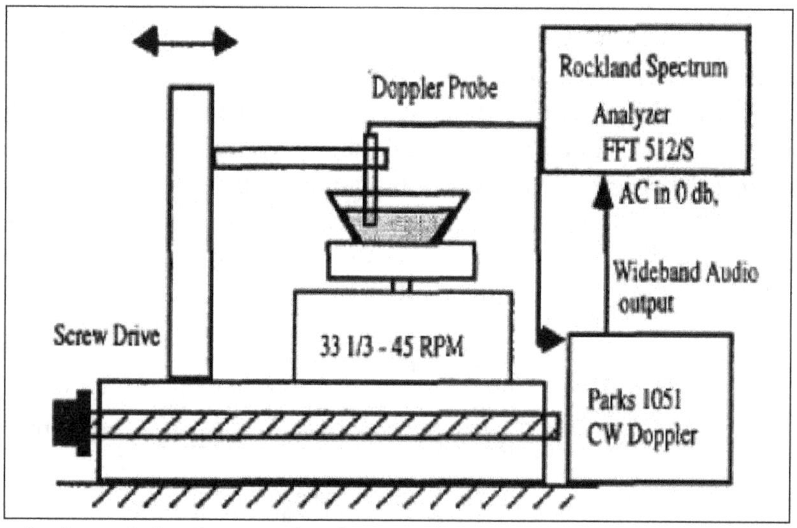

Experimental schematic for CW Doppler ultrasound system for microcirculation measurement.

Although CW system is a convenient technique, the system is unable to resolve the target positions along the direction of propagation of the sound beam through the tissue and is also unable to separate Doppler signals arising from different points along the transmitted ultrasound beam.

Pulsed Wave Doppler Ultrasound Technique

Pulsed Wave (PW) Doppler systems use a transducer that alternates transmission and reception of ultrasound to and from the human tissue. An ultrasound pulse transmitted into the tissue travels for a given time until it is reflected back by a moving red blood cell. It then returns to the transducer over the same time interval but at a shifted frequency. The Doppler shift in frequency is measured and using equation $v_D = \frac{cf_D}{2f_o \mathrm{Cos}\theta}$ the blood flow velocity is calculated.

Transcutaneous pulsed Doppler ultrasound technique was used to measure the blood flow velocity in the Superior Mesenteric Artery (SMA) in 25 healthy newborn infants to investigate the response of intestinal blood flow due to feeding. Duplex scan examination of the SMA was performed in a timed sequence in relation to food intake (fasting, 15-45-90 min following feeding). The Doppler flow signal was analyzed using an integral fast Fourier transform processor and the peak blood velocity (V_p), the pulsatility index (PI) and the mean velocity (Vm) were calculated. The analysis of the flow patterns before and after feeding showed an increase in the V_p and V_m with a corresponding decrease in the PI after feeding. PI refers to the calculation of Doppler measurements of systolic and diastolic blood velocities in the uterine, umbilical or foetal circulations. This technique is rapidly becoming a common clinical procedure in non-invasive neonatal diagnosis. Similar works were carried out in adults to measure Superior Mesenteric Blood Flow (SMBF) in 70 healthy adult subjects of both sexes. The subjects had an overnight fasting and after resting for 30 minutes in a supine position the Doppler ultrasound scanning of the upper abdomen was carried out. From the Doppler shift signals the mean value of SMBF rate was calculated. This method was useful in studying the gastrointestinal blood flow in normal and diseased conditions.

Pulsed ultrasound is also used to assess the arteries located in the base of the brain (Transcranial Doppler Sonography, TDS), to find out any narrowing or blockage in the arteries that may decrease or stop the flow of blood to the brain leading to a stroke. A small hand-held, wand-like probe covered with a conductive gel as shown in figure below will be placed at various areas on the head, usually at each temple, over each eye, and at the base of the skull. The Doppler shift in the frequency was found and the spectral analysis was done from which the blood velocity and the pulsatility index of the flow within the blood vessel were calculated. The painless procedure takes about 20 to 60 minutes after which normal activity can be resumed immediately.

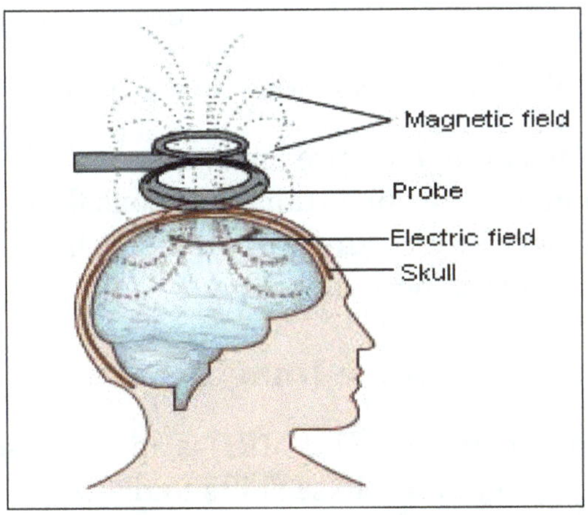

Transcranial Doppler Sonography.

Blood flow velocity in the middle cerebral artery was recorded continuously using the TDS technique during a standard electroencephalogram (EEG) for 30min duration in 51 children with different types of epilepsy. A special holding device incorporating a 2MHz pulsed Doppler probe was fixed by elastic bands to the temporal bone of the skull. Ultrasound intensity of the order of 70-80mW/cm2 was used for the study. The Doppler readings obtained were analysed by fast Fourier real time frequency analysis technique. The results showed a direct relationship of mean blood velocity to the occurrence of tonic seizures/tonic-clonic seizures in the subject which is attributed to the changes in the arterial blood pressure and cerebral metabolism in the presence of distributed autoregulation.

Ultrasound for Detecting 3D Velocity Vector

For the ultrasound systems, only the axial component of the true velocity vector was estimated, whereas the determination of 3D velocity vector was possible by using multiple ultrasound beams. The experiments were carried out using a prototype model (6.5mm diameter dialysis tube) where the blood cell scatterers were tracked as they moved between three ultrasound beams generated by three piezoelectric transducers. The three ultrasound beams are directed and focused by a tilted lens and a parallel, focused and overlapping beam at a desired focal distance is formed. The overlapping of the beam is necessary to ensure that irrespective of the flow direction, the cells will always pass through at least two of the three beam pairs. A 3-D velocity vector algorithm was developed which tracked the movement profile of the cells from beam to beam.

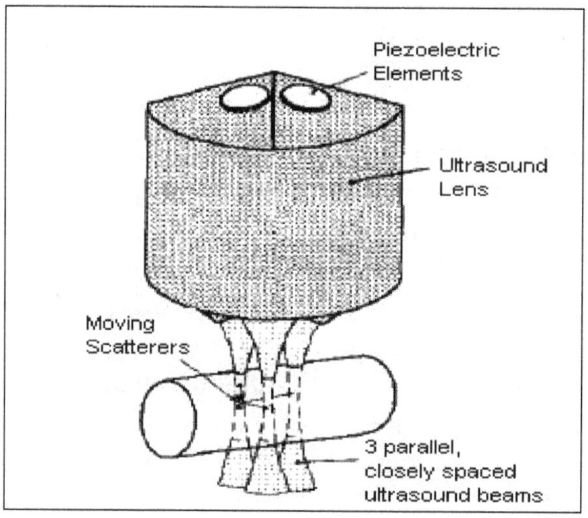

3-D flow velocity vector measurement concept using Ultrasound.

In general, Ultrasound has been used to measure instantaneous value of flow of blood. Ultrasound can be beamed through the skin, thus making transcutaneous flowmeters practical and advanced types of ultrasonic flowmeters can also measure flow profiles. The instrument cannot be calibrated in absolute units but the sensitivity can be standardized.

Phase Contrast Magnetic Resonance Imaging

Phase contrast magnetic resonance imaging (PC-MRI) is another useful method for the measurement of blood flow using velocity induced phase shifts to distinguish between flowing blood and surrounding tissue. A slice of tissue is excited by the application of a RF pulse followed by two equal but opposite bipolar gradients. Protons moving along the direction of the bipolar gradients experience phase shifts, while the stationary tissue does not. The phase shift depends upon how far the flowing protons have travelled between the application of the bipolar gradient pulses. Faster flowing protons will undergo greater phase shifts and hence a measure of the phase shift gives information about the blood flow velocity.

PC-MRI was used to measure the 3D blood flow and the mean velocity in the main, right and left pulmonary arteries of seven healthy volunteers. The moving red blood cells in the blood produce a phase shift of the MRI signal which is used to create images where the pixel intensity is proportional to velocity in each pixel in a single selected direction. The total flow over one cardiac cycle in the main pulmonary artery, right pulmonary artery and left pulmonary artery were estimated.. The axial velocity profiles in the main pulmonary artery showed smaller velocities along the inferior wall of the vessel. The magnitude and directional patterns of flow, assists in understanding phenomena such as thrombus formation (a blood clot formed within a blood vessel and remaining attached to its place of origin), atherosclerosis (fatty deposits inside the arterial walls narrowing the arteries) and changes in endothelial cell structure.

Measurement of blood velocity using PC-MRI technique can also be used for high resolution measurements in the heart or in any part of the circulatory system without the need for catheterization or ionizing radiation. It plays an important role in the evaluation of patients with circulatory disorders. The technique has been tested both with phantoms and on the heart of volunteers using a GE 1.5T scanner. The volume flow per one stroke in the descending aorta and the corresponding cardiac output was measured.

Positron Emission Tomography Imaging

Positron Emission Tomography (PET) is a non-invasive imaging modality for obtaining in-vivo cross sectional images of the human body. In this technique a very small amount of labelled compound (called radiopharmaceutical or radiotracer) is introduced into the patient usually by intravenous injection and after an appropriate uptake period, the concentration of the tracer in the tissue is measured by the scanner. The radionuclide decays by emitting a positron. The emitted positron almost immediately combines with an electron and annihilates each other to emit two gamma rays at 511keV in opposite directions which penetrates the surrounding tissue and are recorded by a circular array of detectors. The spatial distribution of the radioactivity within the subject for a selected plane is reconstructed and the resulting image is displayed on the monitor. Thus, PET provides a non-invasive regional assessment of the biochemical processes necessary for the proper functioning of the organ being imaged.

Regional Myocardial Blood Flow (MBF) was measured accurately and non-invasively using ^{13}N-Ammonia dynamic PET imaging and factor analysis of dynamic structures (FADS) in ten dogs. The MBF estimate results correlated well with the MBF estimates obtained by the two compartment model and microscopic technique and the MBF parametric image had better image quality and lower noise level when compared to those generated by Patlak graphical analysis method.

PET with $H_2{}^{15}O$ was used for measuring regional blood flow to the hind limb skeletal muscle of six anesthetized dogs. Local infusion of adenosine was done to increase blood flow and a hydrooccluder was positioned on the femoral artery to reduce blood flow. The readings obtained by PET imaging were compared with those measured by using radioactive microspheres. The basal flow to the hind limb skeletal muscle was measured as 3.83±0.36ml·min1·100g-1 which was in excellent agreement with the microsphere data given by $3.73\pm0.32 ml\cdot min^{-1}\cdot 100g^{-1}$. The local infusion of adenosine increased the flow to as high as $30 ml\cdot min^{-1}\cdot 100g^{-1}$ and the PET and microsphere data were highly correlated over the entire flow range.

The measurement of blood flow in tumor cells gave information about the uptake of oxygen into the tumor cells in relation to delivery of oxygen through blood and is very important in understanding the pathophysiology and in therapy response.

Role of Optics in Blood Flow Measurement

The advantages of optical imaging are high image resolution, relative low cost (compared to MRI, PET), nonionic, noninvasive and no physical contact. Photoplethysmography is one of the most popular method for measuring the flow volume changes which is based on optical absorption of arterial blood and is most commonly used in current clinical practice. Other important optical techniques for monitoring blood flow are based on the frequency shift of lasers (Doppler effect) and the temporal statistics of time-varying laser speckle. Accordingly they are termed as laser Doppler and laser speckle methods.

Photoplethysmography

The recent advancement in this field is the use of optical techniques in plethysmography, referred to as PhotoPlethysmoGraphy (PPG). The PPG makes use of a laser/light emitting diode(LED) source of selected wavelength (usually in the red/NIR range) to illuminate the tissue and a photodetector to measure the small variations in light intensity associated with changes in perfusion in the catchment volume. PPG has been used to measure microvascular blood flow and tissue viability. The potential of the PPG signal was taken as an objective measure of the healing potential of tissue. If the PPG signal was pulsatile then healing was indicated with conservative management, if non-pulsatile then healing was likely to fail.

The role of the wavelength of the LED(used in the PPG) in perfusion measurement has also been taken up and found successful. The differences in the tissue occlusion pressure has been demonstrated as a function of PPG LED wavelength. This finding helped in developing better methods to predict who might be at risk of getting sores. PPG can also be used to measure the healing potential of ulcers in the skin by measuring the pulsatile cutaneous blood flow. It was found that healing ulcers demonstrated good pulsatile waveforms in cutaneous blood within the immediate vicinity of the ulcer. Cutaneous blood flow around the ulcer was seen to be non-pulsatile if the ulcer did not display evidence of healing. A pulsed PPG operating with different light frequencies made it possible to detect simultaneous flow variations at different levels of dermis/subcutis corresponding to the subpappilary and deep vascular plexi. Despite its simplicity the origin of the different components of the PPG signal are still not fully understood.

Laser Doppler Imaging

Light scattered back from laser irradiated skin provides information on the blood flow in the skin. This is due to the Doppler effect, which causes a frequency shift of light that is scattered by moving red blood cells. This Doppler frequency measured is utilized for the evaluation of the blood flow in the skin. The light from the laser source is delivered to the surface of the skin using a fiber optic cable and the same or a second fiber is used to collect the back scattered light from the skin and returned to a detector. When the incident laser beam is normal to the flow velocity then the Doppler shifted frequency (f) is given by the equation below where v is the flow velocity, λ is the wavelength of the laser beam and θ is the angle between the incident and the scattered beam. Once the Doppler shift frequency is measured then the velocity can be calculated.

$$f = \frac{v.\sin\theta}{\lambda}$$

A laser Doppler velocimeter was first used to measure the blood velocity in retinal arteries of albino rabbits. A He-Ne laser with an output power of 10mW was used and the exposure time for one measurement was 2minutes. The magnitude of the Doppler shifted frequency was measured by optical heterodyning using a photo multiplier tube which is followed by an amplifier, a spectrum analyzer, a squarer and an integrator. The blood flow velocities measured were in the range of 1.1 to 1.8cm/sec. A fundus camera laser Doppler velocimeter has also been used to measure the relative blood velocity in the human optic nerve head. The maximum blood velocity measured was in the range of 0.12 to 0.2 mm/sec.

In-vivo blood flow measurement was conducted on the dissected lateral bony wall of cochlea of normal guinea pigs using a Laser Doppler Flowmeter (LDF). The output signal of laser Doppler flowmeter was recorded on a strip chart recorder. The effect of the drug norepinephrine administered from the femoral vein was studied without and after placing a dissected piece of bone on the lateral wall of the cochlea. The LDF signal patterns obtained were similar in both the cases but the magnitude of the output signal decreased to about one fourth the original signal. Placement of the bone reduced the magnitude of the light penetrating through the perfused tissue and as a result the blood volume detected had decreased.

In another reported work, the bronchial mucosal blood flow was measured in pigs using LDF. The source was a 2mW He-Ne laser with a wavelength of 632.8nm.The viability of the bronchial anastomosis after lung transplantation depends on blood flow from the pulmonary circulation to the bronchial circulation, during the first days after surgery. The standard left thoractomy was made and the bronchial mucosal blood flow was measured, when pulmonary artery, bronchial artery, or both were clamped. The measured bronchial mucosal blood flow diminished by one-third when the pulmonary artery was clamped and by two-thirds when the bronchial artery was clamped which showed that bronchial mucosal blood flow after division of the bronchial circulation is about onethird of the normal value. When both bronchial and pulmonary circulations were closed, a small amount of perfusion could still be measured which was due to the sensitivity of the technique.

A laser Doppler scanner technique was used for blood flow imaging in the capillaries under the surface of the skin. Instead of using optical fibres, the laser beam from a 2mW He-Ne laser was moved in a raster pattern over the skin to build up a laser Doppler image the advantage being that the scanner cannot influence the blood flow. The scanning speed was set at a normal speed of 1s per line and low speed of 9s per line. Experiments were conducted on human subjects and the laser Doppler images of the hand and chest of normal subjects were obtained. Using the technique an image of blood flux in the capillaries under the surface of the skin was obtained and it was also possible to detect changes in blood flux resulting from injury, disease and temperature variations.

In another study two Doppler blood flow measuring heads each consisting of a diode laser of 5mW, with a wavelength of 685nm or 785nm was used to irradiate the skin tissue and the reemitted light was sampled using two photodiodes placed at a distance of 12mm and at an angle of 450 to the skin surface. The measuring heads were connected to a controller card in the PC and using a fast Fourier transform the frequency spectra was obtained which was then converted into a flow spectrum. Upto four frequency ranges for normalized blood flows could be displayed simultaneously in different colors. The blood flow measured showed different behavior for low and high Doppler frequencies and thus provided the possibility to distinguish between the flow in the superficial dermal plexus and larger micro-capillaries.

LDF has also been used to study the optic nerve head (ONH) blood flow in the eye which responds to increased neuronal activity induced by visual stimulation. A fundus camera was modified by replacing the light source (lamps and associated optics) of the camera by 12 light emitting diodes radiating at 770±20nm and a near-infrared laser diode lasing at 810nm was focused on the optic nerve tissue. An optical fibre collected the light from the site illuminated by the laser and an avalanche photodiode was used to convert the light into electric current which was then amplified, digitized and fed to the computer. The blood flow response to diffuse stimulation was recorded. Using the Doppler shift power spectrum the blood velocity, flow and volume were calculated. The instrument was found to be useful in the study of the role of the blood circulation in the pathogenesis of various ocular diseases of vascular origin, such as glaucoma and diabetic retinopathy.

The laser Doppler imaging technique has the advantages of being a non-invasive simple to apply technique, providing a continuous or near continuous record but suffers from the disadvantages of limited laser penetration depth, low resolution, longer measurement time, not being a whole field and real time technique and the measurement values are not absolute but relative in nature.

The sensitivity of the instrument to a particular blood flow rate in a blood vessel varies with the distance of the vessel from the probe tip and the distribution of blood flow in a tissue through its different layers will not remain constant under different conditions. It was therefore not possible to calibrate laser Doppler flowmeters in absolute units.

Laser Speckle Contrast Imaging

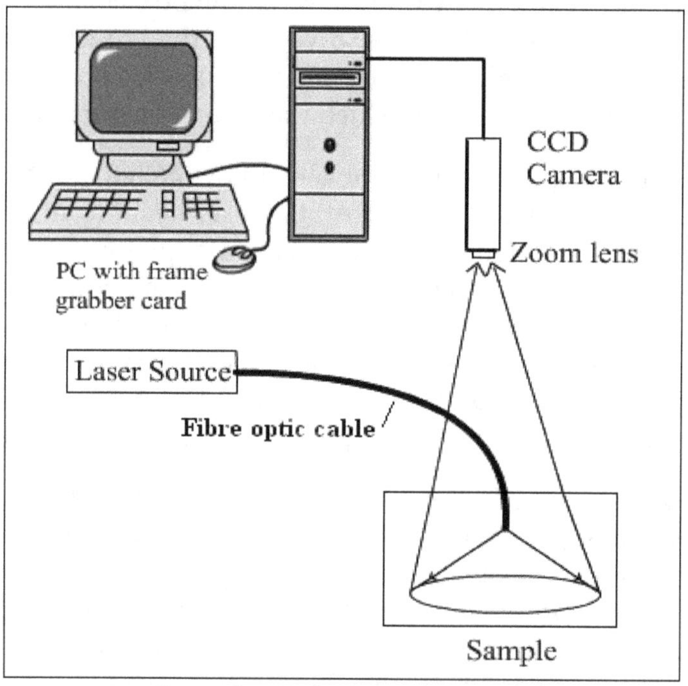

Experimental setup for laser speckle imaging.

The laser speckle technique, over and above the laser Doppler technique offers the advantages of being a real time, non-scanning whole field technique offering higher resolution. The schematic diagram of the experimental setup shown in figure below requires only standard hardware components. The light from a laser source illuminates the object of interest. A CCD (Charge Coupled Device) camera images the illuminated area and the speckle pattern is captured by the frame grabber and digitized, then processed on the computer and displayed on the PC monitor. When fairly coherent light is either reflected from a rough surface or propagates through a medium with random refractive index fluctuations a random intensity distribution pattern is formed which is called as a laser speckle pattern which can be explained only by statistics. The statistical properties of speckle patterns depends both on the coherence of the incident light and the detailed properties of the random surface or medium. In the early days of lasers, the effect was regarded purely as a nuisance, but later scientists started to study speckle and developed practical applications of the phenomenon.

The intensity of the speckle pattern is not deterministic and can only be described by statistics. Utilising the first order spatial statistics, the property of a fully developed speckle pattern is that the standard deviation of the spatial intensity variations is equal to the mean intensity. In practice, the speckle patterns have a standard deviation that is less than the mean intensity, and hence there is a corresponding reduction in the contrast of the speckle pattern. The expression for the speckle contrast of a speckle pattern is given by equation:

$$\text{Speckle contrast}\, K_s = \frac{\sigma_s}{\langle I \rangle}$$

where σ_S is the standard deviation of the spatial intensity variations in the speckle pattern and $\langle I \rangle$ is the spatial average of the intensity fluctuations. Hence, for a fully developed speckle pattern the contrast is defined to be unity and for a partially developed speckle pattern the speckle contrast is less than 1.

An ideal speckle pattern will have a probability density function of intensity that is negative exponential in nature. It can be shown that the standard deviation of the intensity variations in an ideal speckle pattern is equal to the mean intensity. Thus the contrast of a polarised speckle pattern is always unity. But in practice, a speckle pattern often has a standard deviation less than the mean intensity and this is observed as a reduction in the contrast of the speckle pattern.

Goodman related the variance of the spatial fluctuations to the time average of the auto-covariance, $C_\tau(\tau)$ of the intensity fluctuations given by equation,

$$\sigma_s^2(T) = \frac{1}{T}\int_0^T C_\tau(\tau)\,d\tau$$

If we assume that it has a Lorentzian velocity distribution (because the capillary network is so convoluted) and no multiple scattering and perfect speckle formation then the speckle contrast is given by equation below where, T is the exposure time and τ_c is the correlation time of the intensity fluctuations (i.e. the time taken for intensity to fall to 1/e the original value).

$$K_s = \sqrt{\frac{\tau_c}{2T}\left\{1 - \exp\left(\frac{-2T}{\tau_c}\right)\right\}}$$

The simplest approach defines the de-correlation velocity v_c as given by equation:

$$v_c = \frac{\lambda}{2\pi\tau_c}$$

The blood flow can be calculated if the area of illumination is known. The flow of blood in a vessel is related to velocity by the equation below where F = blood flow, v_c = mean velocity, and A = cross-sectional area of the blood vessel.

$$F = v_c \times A$$

When an object moves, the speckle pattern also changes. For small movements the speckles move with the object and remain correlated. For larger motions the speckles de-correlate and the speckle pattern changes completely. The de-correlation occurs as light is scattered from a large number of individual scatterers that are moving in a fluid. The moving speckle pattern contains information about the object's motion. One of the most important applications of this theory is to study the fluctuations caused by the flow of blood.

The frequency spectrum of the fluctuations is dependent on the velocity of blood. The state of the art in this technique begins with a single exposure speckle photography using a 50mW He-Ne laser source for imaging the human retinal blood flow carried out by J.D. Briers group in the early 1980s. The basic underlying principle was that the speckle pattern in an area where flow occurs will be blurred to an extent which will depend on the velocity of flow and on the exposure time of the photograph whereas in an area of no flow the speckle pattern is fully developed. The velocity distribution in the flow area is mapped as a variation in contrast of the speckle pattern. A high pass optical filtering technique was adopted to convert the contrast variations into intensity variations and differences in the speckle contrast between blood vessels and surrounding areas could be visualized. The disadvantage of this method is that it was a two stage process that precluded real time operation. As a clinical technique the method was not viable.

A modified method of monitoring blood flow using optical fibres and photomultiplier was proposed and in such an attempt, a He-Ne laser was used to illuminate the fingertip of human subjects. The intensity fluctuations were observed at a certain point and was detected through a pinhole by a photomultiplier, then amplified and fed to a spectrum analyzer. The signal was plotted for normal flow and reduced flow by placing an inflated cuff on the upper arm. The analysis of the power spectral distribution (PSD) of the speckle pattern gave useful information on skin blood flow. Experiments were also conducted to study the use of optical fibres with variation in diameter for illumination and to obtain the dynamic speckle patterns for velocity determination. The same group also investigated influence of illumination fiber movement, object movement and diameter of the detecting aperture on the time dependent behavior of the speckle pattern.

In a contemporary study, the speckle signal from the skin of human subjects was detected using a CCD image sensor and a colour graphic display of the 2D microcirculation map was developed. A He-Ne laser line spot was used to illuminate the skin surface and the output pulses from the CCD image sensor was digitized using an analog to digital converter and stored in a static RAM and the data was transferred to a host microcomputer. The difference in data between successive scans was calculated and integrated for each pixel point and the blood flow map was obtained. When a rubber band was placed on one of the fingers there was a corresponding decrease in the blood flow.

Blood flow evaluations have also been carried out in the gastric mucous membrane of anaesthetized albino rabbit using bio-speckles (speckles obtained from living objects). The power spectral distributions (frequency in kHz versus power spectrum in dB) were obtained for normal, reduced and enhanced blood flow. A relative increase of the lower and higher frequency components was observed in the power spectral distribution in response to reduced and enhanced blood flow.

A quasi real-time digital version of the single exposure speckle photography technique was developed. A CCD camera and a frame grabber card were used to obtain the image of the hand the experimental setup of which is as shown in figure, and the processing time was about 4 minutes. The drawback of the technique was the need to reduce the processing time and to introduce the option to vary the exposure time so that the velocity range being measured can be varied.

Following this, a new non-invasive, fully digital and real time technique for monitoring blood flow termed as LAser Speckle Contrast Analysis(LASCA) based on the first order spatial statistics of time integrated speckle was developed. It was a fully digital real time version of the single exposure speckle photography technique. The experimental setup used included a CCD camera, frame grabber card and specially developed software which was used to compute the speckle contrast of capillary blood flow. The contrast variation was then converted to velocity. Experiments were conducted to study the effect of temperature on blood flow. Human subjects were asked to immerse one hand in cold water (3 to 100C for 30s) and the other in warm water (41 to 480C for 2min). LASCA was used to calculate the speckle contrast variations in an area on the back of the hand. For all subjects the contrast measured on the warm hand was lower than that measured on the cold hand. The changes in speckle contrast after conversion to blood velocity showed an increase of velocity of about 10% to 50%.

The LASCA technique has also been used to produce 2D velocity maps of blood flow in capillaries. The processed image of 3 fingers of a subject was obtained at different exposure times. Higher velocity values obtained using the faster exposure times suggested that the blood flow is being measured from deeper blood vessels in the finger. Trials were also done using red laser to grab a series of images which could then be post-processed to produce a video clip of the changes in blood flow. Although similar experiments carried out using a green laser (514nm) couldn't give substantial information regarding the blood flow parameters, the use of green laser for illumination may provide a method for removing the artifacts due to gross body movements. The effects due to body movements could also be corrected by placing an opaque scatterer attached to the patient's skin or by applying suitable algorithms. Using one such algorithm the processing time was brought down to few seconds from several minutes.

Spatial statistics of time-integrated speckle was used to obtain the image of the hand illuminated by laser light. Modifications were introduced in the existing laser speckle imaging (LSI) method, and based on the temporal statistics a time-integrated speckle was developed. Cerebral blood flow map under different temperatures was measured on the rat cortex. The spatial resolution of the modified LSI method was five times higher compared to LASCA method and more blood vessels appeared clearly in the image. When the temperature was increased from 35 to 500C the blood flow map obtained showed an increase in blood perfusion. The disadvantage of this method was that it was an invasive method and the temporal resolution was also much lower when compared to LASCA method but however, this method could be used to measure relatively slow changes of blood flow.

Blood velocity measurements were made on the elevated flaps of the skin of juvenile pigs using LASCA/MS multiple scattering method. The readings obtained were compared to those obtained using laser Doppler velocimetry technique and there was a good correlation between the two. The same principle was also used to study the retinal blood velocity in rabbits using a modified fundus camera in which the pulsed light source was replaced using a CW He-Ne laser.

The cortical vascular blood flow pattern of rats which is associated with the functional response in cerebral cortex was obtained using temporal LASCA method. A centerline extraction method based on intensity ridge tracking was used to segment the blood vessels from the acquired raw speckle image and the blood vessel patterns could be clearly seen. Although the technique is an invasive method, the vascular blood flow pattern obtained could be used to study the spatiotemporal activities of the somatosensory centre and in the diagnosis of the focal stroke or ischemia (inadequate blood supply (circulation) to a local area due to blockage of the blood vessels to the area). A simplified LASCA imaging analysis method which improved the computation time was developed. The method was applied to image blood flow in the retina of three rats under normal and physiologic challenge conditions. The method may be applied for the early detection of retinal diseases.

Another group which was motivated by the usefulness of laser Doppler perfusion images and the advantages of full field illumination using a laser speckle imaging approach went in for designing a new hybrid technique which they called as laser speckle perfusion imaging (LSPI). They designed an in vitro blood flow model to mimic a range of tissue optical characteristics. The model simulated a range of red blood cell concentrations and velocities in a matrix with absortive and scattering properties similar to tissues of interest. The results obtained validated the use of the technique for measuring tissue perfusion.

Monitoring cerebral blood flow (CBF) is important for studying both normal and pathophysiologic conditions of brain metabolism. Many groups around the world have made use of LASCA for full field imaging of cerebral blood flow dynamics in animal models. Monitoring blood flow dynamics in the in vivo rodent dorsal skin fold model using laser speckle imaging has been carried out. A new image processing method for laser speckle, using adaptive windowing laser speckle contrast analysis (aLASCA) that adaptively processes laser speckle images in the temporal and spatial direction has been developed. The most recent application of LASCA was to conduct experiments on the retina of male Long-Evans rats in which laser speckle flowmetry (LSF) was used in conjunction with confocal microscopy to monitor light-evoked changes in blood flow in retinal vessels. This dual imaging technique aided the group to stimulate retinal photoreceptors and measure vessel diameter with confocal microscopy while simultaneously monitoring blood flow with LSF. LASCA has also started giving promising results in analyzing real time changes in blood flow associated with different stages of skin diseases.

Cardiac Pacemakers

The human heart beats rhythmically at the rate between 60-100 beats/minute. Ever wondered what decides the rate at which our heart keeps beating? It is in the Sinoatrial node of the heart, where there are cells which generate these impulses. As they set the pace for beating of heart, they are appropriately called pacemaker cells, as they directly control the heart rate. They make up the cardiac pacemaker, that is, the natural pacemaker of the heart.

Sometimes, when there is failure of the natural pacemaker of the heart or there is a block in the conduction system not allowing the impulse to conduct, there is a need for an artificial pacemaker. An artificial pacemaker is a device which uses electrical impulses, delivered by electrodes to the heart muscle and therefore regulates the beating of the heart. The chief purpose of an artificial cardiac pacemaker is to maintain a suitable heart rate.

New generation pacemakers, generally also have several functions besides this. The most elementary ones monitor the heart's inherent electrical rhythm. And whenever there is a failure to detect a heartbeat within a stipulated time the pacemaker stimulates the ventricle of the heart with a short low voltage pulse. Both sensing and stimulating take place on a beat to beat basis. The more complex ones include the ability to sense and/or stimulate both the atrial and ventricular chambers.

Physiology of Pacemaker of Heart

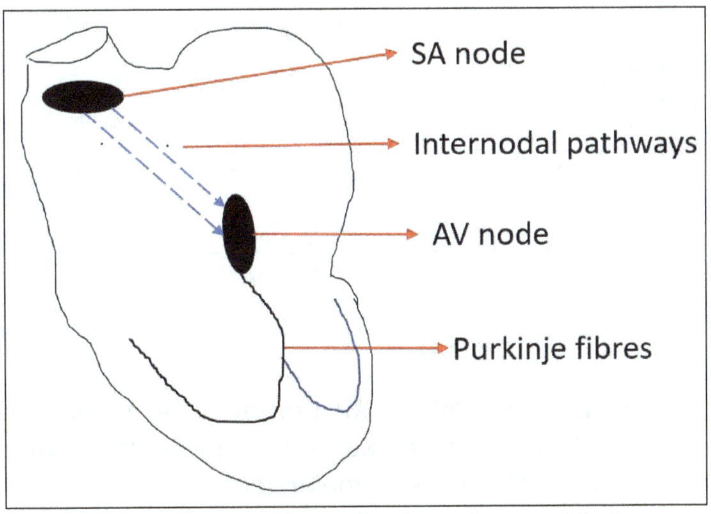

Simplified diagram of the conducting pathways of the heart.

There are two properties of the SA node cells which make them appropriate to be the pace maker of the heart: automaticity and rhythmicity. Automaticity is the property of a tissue to generate impulses spontaneously while, rhythmicity means that the pulse is generated at regular intervals.

Conducting System of the Heart

The conduction of the cardiac impulse begins at the sinoatrial node lying in the superior posterolateral wall of the right atrium from where it conducts to atrioventricular node and finally spreads to ventricular musculature. The rate at which different parts of this conduction system discharge impulses is different. The fastest impulse generator is the SA node followed by the AV node. Being the fastest, it is the rate at which the SA node fires which decides the rate or pace of the heart. This is so because an impulse originating in the SA node conducts through the conduction system and causes serial contraction of different chambers of the heart. The conducting system of the heart consists of specialized fibres of the heart muscle present as the sinoatrial node, the internodal tracts, the atrioventricular (AV) node, the bundle of His and its right and left terminal branches, and the Purkinje fibres. Before an impulse from the AV node or any other part of the conduction system can initiate an impulse, the next impulse from the SA node is discharged because its rate of firing is the highest.

Electrical Activity of the Nodal Tissue

The nodal tissue is unique as it does not have a constant 'resting membrane potential' which is the difference in electrical potential across the cell membrane seen at rest. Although the tissue is at rest this potential in the nodal tissue moves towards depolarisation. And once the threshold is achieved an impulse called the action potential is fired. After firing of this impulse, the potential comes back to the resting levels. This instability of the resting membrane potential in the cardiac tissue generates its automaticity and the repetition of this cycle causes its rhythmicity. For the SA node this rate is around 70/minute. The slow depolarisation or rise of the potential during the resting phase is called pacemaker potential.

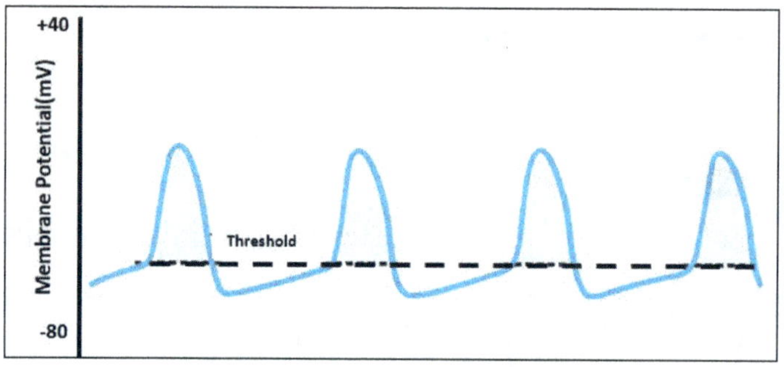

Rhythmic electrical activity in the SA node.

Ionic Basis of Pacemaker Potential

- Leakiness of the cell membrane: The cell membrane of the nodal tissue is relatively leaky or permeable to ions at all the times. Its maximum leakiness is to sodium ions. It is the inward leak of sodium ions inwards that cause depolarization.
- Funny current: It is generated by both sodium and potassium movement and is mainly responsible for the generation of pacemaker potential. It has a threshold of -40 to -50 millivolts and becomes more as polarisation increases.
- Contribution of calcium: Calcium ions also contribute to the pacemaker potential. Inward flow of calcium increases depolarisation, making it reach the threshold faster.

Disorders of the Conduction System

Under physiological conditions, the conduction system of the hear allows an impulse generated by depolarisation of the SA node to propagate through the heart causing its rhythmic beating. The electrical conduction of the impulse causes the mechanical beating of the heart. There are two main disorders of the conduction system which can adversely impact the heart's electrical system.

- Bradycardia, when the heart beats too slowly and patient will present with the symptoms of fatigue, dizziness, fainting, and lethargy.

- Tachycardia, when the heart beats too fast and the patient presents with palpitations, racing heart and pounding in the chest.

Both these disorders are due to three main reasons:

- First, are SA node disorders, where the SA node fails to generate electrical impulses properly. This will cause the patient's heart to beat slowly.
- Second, there are various grades of heart block in which the conducting pathway of the heart is defective at the AV node. Here, the SA node is functioning correctly, but the impulse doesn't get relayed from the AV node, as a result there is bradycardia. The heart's pumping becomes inefficient and the patient feels easily tired, out of breath, dizzy, and uncomfortable.
- In the third situation, there are aberrant conduction pathways causing the impulse to circle back into the conducting pathways stimulating the same cardiac cells repeatedly. Instead of progressing through the heart and dissipating, this re-entry loop allows one signal to cause the heart to keep beating faster causing tachycardia.

Indications for Artificial Pacemakers

Cardiac pacemakers are implanted most commonly in patients who have a heartbeat slower than usual called bradycardia. But, more important than bradycardia are the symptoms that may arise because of it, irrespective of whether the bradycardia has been documented or not. Pacing can be done to prevent symptomatic bradycardia in asymptomatic patients if the risk for rapid development to serious symptoms is there.

Types of Permanent Pacemakers

- Single-chamber pacemaker: One pacing lead is placed into a chamber of the heart, either the atrium or the ventricle.
- Dual-chamber pacemaker: Leads are placed in two chambers of the heart. One lead paces the atrium and one paces the ventricle. This type more closely resembles the natural pacing of the heart by assisting the heart in coordinating the function between the atria and ventricles.
- Rate-responsive pacemaker: Here the changes in the patient's physical activity are detected and pacing rate automatically adjusted to fulfil the body's needs.

Components of a Cardiac Pacemaker

The pacemaker system consists of three components:

- Pulse generator, which is an implantable device.
- Lead that connect this pacemaker system to the heart.
- A computer programmer to communicate with the implanted device.

The main component of this pacemaker system is the pulse generator and therefore it is often referred to as the pacemaker. But, it is a requirement that all the three components work in synchrony for successful pacing.

Functioning of the Pacemaker System

The essential requirement from a pace making device is that it delivers precisely timed small output electrical pulses. Hence, there are two vital things which this device requires: first, is the requirement of a timer and second, is a power source for delivering the pulses. Besides this, the present generation of pacemakers also contains a complex circuitry which can sense the incoming information and then respond by giving pulses appropriately. Some of the newer devices contain memory and can store electrocardiograms which can be downloaded by the clinician.

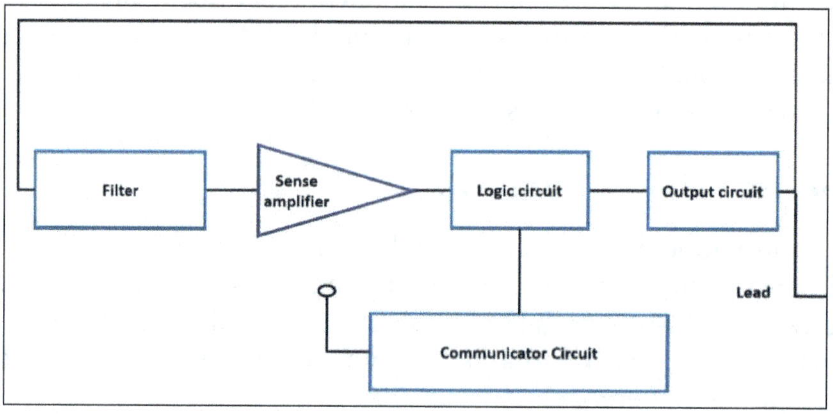

Simplified circuit diagram of a pacemaker.

Timing Mechanism

Pacemakers receive inputs from the leads implanted in the heart and functioning as a sensor. These sensors, sense the electrical activity of the heart and enables the device to pace according to the requirement of the heart.

A problem with the sensing of electrical activity is that the body essentially generates a lot of electricity and it is commonly seen that the activity of the ventricles generally drowns the signal generated by the smaller atria. A way of removing this ventricle noise is to filter the signal.

Filtering enables us to find the signal of choice from the electrical activity is that a recorded. The clinician adjusts this filter settings, making the device less or more sensitive. Although, the final signal that we get maybe weak and hence a sense amplifier is installed to boost the signal making it large enough for the device to work with. A comparator compares the filtered and the amplified signal to baseline after which the signal enters the logic circuitry of the device, and is evaluated.

Energy Source

The second requirement is of an energy source. For this a battery (in a titanium case) is present in every pacemaker, and takes up almost 50% of the space. Pacemakers usually require around 2.8

volts but since pacemakers must deliver greater than this, we depend on voltage amplifier which is the circuit used to increase the voltage output to increase the delivered voltage. This amplifier will store the energy till the amount available is appropriate for delivering the pulse.

These days, lithium-iodide battery is used and many of the pacemakers also report the battery status on the programmer. One of the greatest benefits of a lithium iodide battery is that it's life span is predictable, therefore the course (of replacement) is easily predictable.

Lifespan of Pacemakers and Pacing Threshold

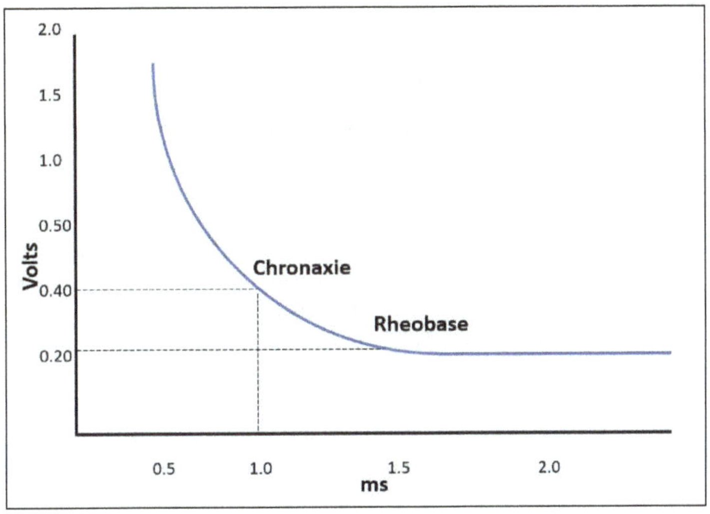

The strength–duration curve depicting the amount of energy, it takes to capture the heart. Rheobase is the point at which the plateau begins and roughly establishes the minimum voltage requirements to capture the heart. The point at which twice the rheobase voltage value meets the curve is defined as chronaxie.

The actual life span of a pulse generator will depend not only on the battery but its usage too. Every battery will require some level of housekeeping current and even if the pulse generator doesn't deliver any pulse, eventually its battery will wear out. The most important factor, deciding device longevity is the output pulse. The two variables which are modifiable are the pulse amplitude (measured in volts) and the pulse width or pulse duration (measured in milliseconds).

The pacing threshold is the lowest amount of energy that is required to consistently cause depolarization of the heart. Representation of this is the strength–duration curve, with rheobase defining the minimum voltage needed to capture the heart and chronaxie (two times rheobase voltage, extended to meet the curve) as a reasonable safety margin. The safety margin is the amount added to the pacing threshold for programming a pacemaker output to assure that the pacemaker will capture the heart, even if the pacing threshold changes. As a rule, the cardiologist determines the pacing threshold and then doubles it.

The most common pacing problems can have been classified as:

- Failure to capture where pacing output is below threshold.

- Failure to pace, may be due to over-sensing of physiologic or non-physiologic signals.
- Pacing at a rate inconsistent with the programmed rate, due to problem with sensing.

Other 'Pacemaker Like' Devices

Resembling pacemakers are other devices called implantable cardioverter-defibrillators (ICDs). These devices can be life saving for patients at risk for sudden cardiac death. An ICD has the ability to treat many types of heart rhythm disturbances by pacing, cardioversion, or defibrillation.

References

- Electromyography: britannica.com, Retrieved 07, June 2020
- Measuring-blood-flow-techniques-and-applications-a-review: researchgate.net, Retrieved 30, January 2020

5

Diverse Aspects of Biophysics

The various topics studied in association to biophysical structure and phenomena are enzyme kinetics, molecular motor, myoglobin, hemoglobin, pancreatic ribonuclease, antibody, collagen and transfer RNA. In order to completely understand biophysical structure and phenomena it is necessary to understand the processes related to it. The following chapter elucidates these varied processes and mechanisms associated with this area of study.

Enzyme Kinetics

Dihydrofolate reductase from *E. coli* with its two substrates dihydrofolate (right) and NADPH (left), bound in the active site. The protein is shown as a ribbon diagram, with alpha helices in red, beta sheets in yellow and loops in blue. Generated from 7DFR.

Enzyme kinetics is the study of the chemical reactions that are catalysed by enzymes. In enzyme kinetics, the reaction rate is measured and the effects of varying the conditions of the reaction are investigated. Studying an enzyme's kinetics in this way can reveal the catalytic mechanism of this enzyme, its role in metabolism, how its activity is controlled, and how a drug or an agonist might inhibit the enzyme.

Enzymes are usually proteinmolecules that manipulate other molecules — the enzymes' substrates. These target molecules bind to an enzyme's active site and are transformed into products through a series of steps known as the enzymatic mechanism.

$$E + S \rightleftarrows ES \rightleftarrows ES^* \rightleftarrows EP \rightleftarrows E + P$$

These mechanisms can be divided into single-substrate and multiple-substrate mechanisms. Kinetic studies on enzymes that only bind one substrate, such as triosephosphate isomerase, aim to measure the affinity with which the enzyme binds this substrate and the turnover rate. Some other examples of enzymes are phosphofructokinase and hexokinase, both of which are important for cellular respiration (glycolysis).

When enzymes bind multiple substrates, such as dihydrofolate reductase, enzyme kinetics can also show the sequence in which these substrates bind and the sequence in which products are released. An example of enzymes that bind a single substrate and release multiple products are proteases, which cleave one protein substrate into two polypeptide products. Others join two substrates together, such as DNA polymerase linking a nucleotide to DNA. Although these mechanisms are often a complex series of steps, there is typically one *rate-determining step* that determines the overall kinetics. This rate-determining step may be a chemical reaction or a conformational change of the enzyme or substrates, such as those involved in the release of product(s) from the enzyme.

Knowledge of the enzyme's structure is helpful in interpreting kinetic data. For example, the structure can suggest how substrates and products bind during catalysis; what changes occur during the reaction; and even the role of particular amino acid residues in the mechanism. Some enzymes change shape significantly during the mechanism; in such cases, it is helpful to determine the enzyme structure with and without bound substrate analogues that do not undergo the enzymatic reaction.

Not all biological catalysts are protein enzymes; RNA-based catalysts such as ribozymes and ribosomes are essential to many cellular functions, such as RNA splicing and translation. The main difference between ribozymes and enzymes is that RNA catalysts are composed of nucleotides, whereas enzymes are composed of amino acids. Ribozymes also perform a more limited set of reactions, although their reaction mechanisms and kinetics can be analysed and classified by the same methods.

General Principles

The reaction catalysed by an enzyme uses exactly the same reactants and produces exactly the same products as the uncatalysed reaction. Like other catalysts, enzymes do not alter the position of equilibrium between substrates and products. However, unlike uncatalysed chemical reactions, enzyme-catalysed reactions display saturation kinetics. For a given enzyme concentration and for relatively low substrate concentrations, the reaction rate increases linearly with substrate concentration; the enzyme molecules are largely free to catalyse the reaction, and increasing substrate concentration means an increasing rate at which the enzyme and substrate molecules encounter one another. However, at relatively high substrate concentrations, the reaction rate asymptotically approaches the theoretical maximum; the enzyme active sites are almost all occupied and the reaction rate is determined by the intrinsic turnover rate of the enzyme. The substrate concentration midway between these two limiting cases is denoted by K_M.

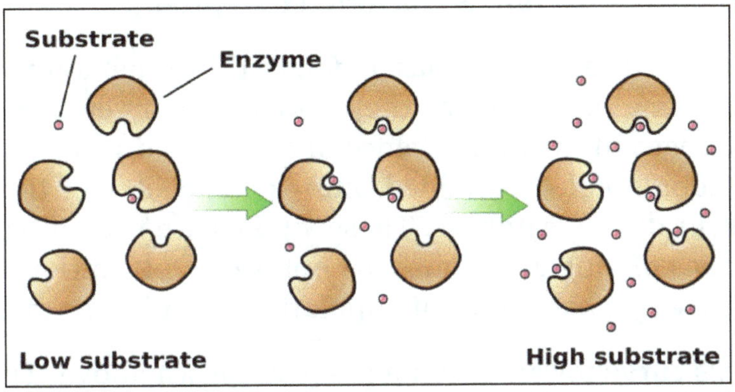

As larger amounts of substrate are added to a reaction, the available enzyme binding sites become filled to the limit of V_{max}. Beyond this limit the enzyme is saturated with substrate and the reaction rate ceases to increase.

The two most important kinetic properties of an enzyme are how quickly the enzyme becomes saturated with a particular substrate, and the maximum rate it can achieve. Knowing these properties suggests what an enzyme might do in the cell and can show how the enzyme will respond to changes in these conditions.

Enzyme Assays

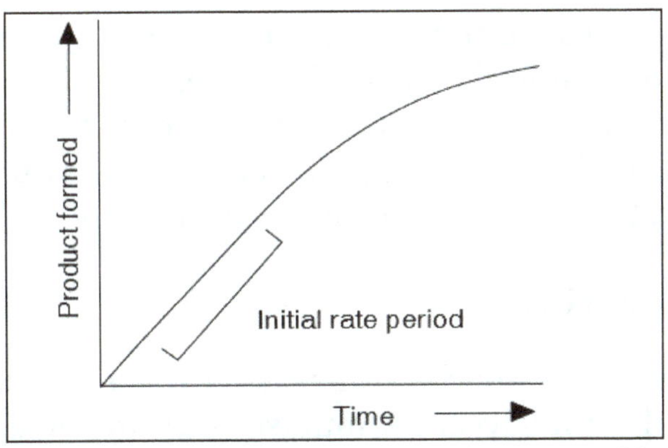

Progress curve for an enzyme reaction. The slope in the initial rate period is the initial rate of reactionv. The Michaelis–Menten equation describes how this slope varies with the concentration of substrate.

Enzyme assays are laboratory procedures that measure the rate of enzyme reactions. Because enzymes are not consumed by the reactions they catalyse, enzyme assays usually follow changes in the concentration of either substrates or products to measure the rate of reaction. There are many methods of measurement. Spectrophotometric assays observe change in the absorbance of light between products and reactants; radiometric assays involve the incorporation or release of radioactivity to measure the amount of product made over time. Spectrophotometric assays are most convenient since they allow the rate of the reaction to be measured continuously. Although radiometric assays require the removal and counting of samples (i.e., they are discontinuous assays) they are usually extremely sensitive and can measure very low levels of enzyme activity. An analogous approach is to use mass spectrometry to monitor the incorporation or release of stable isotopes as substrate is converted into product.

The most sensitive enzyme assays use lasers focused through a microscope to observe changes in single enzyme molecules as they catalyse their reactions. These measurements either use changes in the fluorescence of cofactors during an enzyme's reaction mechanism, or of fluorescent dyes added onto specific sites of the protein to report movements that occur during catalysis. These studies are providing a new view of the kinetics and dynamics of single enzymes, as opposed to traditional enzyme kinetics, which observes the average behaviour of populations of millions of enzyme molecules.

An example progress curve for an enzyme assay is shown above. The enzyme produces product at an initial rate that is approximately linear for a short period after the start of the reaction. As the reaction proceeds and substrate is consumed, the rate continuously slows (so long as substrate is not still at saturating levels). To measure the initial (and maximal) rate, enzyme assays are typically carried out while the reaction has progressed only a few percent towards total completion. The length of the initial rate period depends on the assay conditions and can range from milliseconds to hours. However, equipment for rapidly mixing liquids allows fast kinetic measurements on initial rates of less than one second. These very rapid assays are essential for measuring pre-steady-state kinetics, which are discussed below.

Most enzyme kinetics studies concentrate on this initial, approximately linear part of enzyme reactions. However, it is also possible to measure the complete reaction curve and fit this data to a non-linear rate equation. This way of measuring enzyme reactions is called progress-curve analysis. This approach is useful as an alternative to rapid kinetics when the initial rate is too fast to measure accurately.

Single-substrate Reactions

Enzymes with single-substrate mechanisms include isomerases such as triosephosphateisomerase or bisphosphoglycerate mutase, intramolecular lyases such as adenylate cyclase and the hammerhead ribozyme, an RNA lyase. However, some enzymes that only have a single substrate do not fall into this category of mechanisms. Catalase is an example of this, as the enzyme reacts with a first molecule of hydrogen peroxide substrate, becomes oxidised and is then reduced by a second molecule of substrate. Although a single substrate is involved, the existence of a modified enzyme intermediate means that the mechanism of catalase is actually a ping–pong mechanism, a type of mechanism that is discussed in the *Multi-substrate reactions* section below.

Michaelis–Menten Kinetics

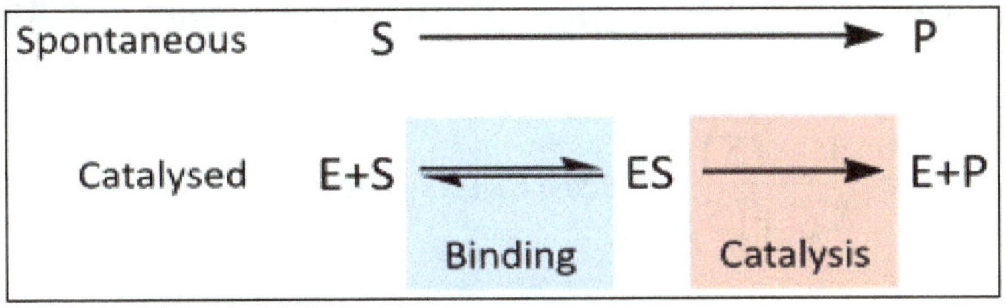

A chemical reaction mechanism with or without enzyme catalysis. The enzyme (E) binds substrate (S) to produce product (P).

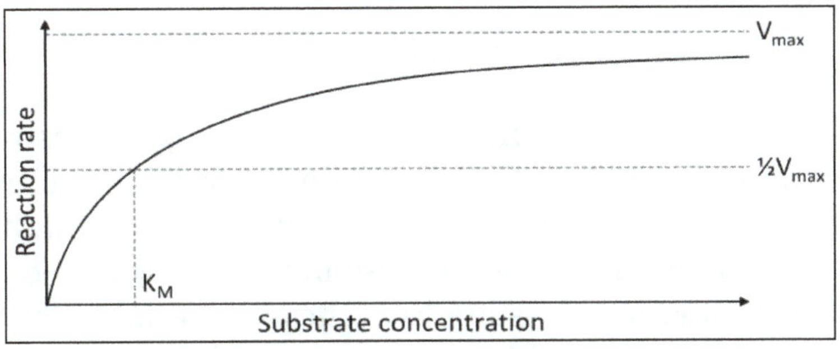

Saturation curve for an enzyme reaction showing the relation between the substrate concentration and reaction rate.

As enzyme-catalysed reactions are saturable, their rate of catalysis does not show a linear response to increasing substrate. If the initial rate of the reaction is measured over a range of substrate concentrations (denoted as [S]), the reaction rate (v) increases as [S] increases, as shown on the right. However, as [S] gets higher, the enzyme becomes saturated with substrate and the rate reaches V_{max}, the enzyme's maximum rate.

The Michaelis–Menten kinetic model of a single-substrate reaction is shown on the right. There is an initial bimolecular reaction between the enzyme E and substrate S to form the enzyme–substrate complex ES. The rate of enzymatic reaction increases with the increase of the substrate concentration up to a certain level called V_{max}; at V_{max}, increase in substrate concentration does not cause any increase in reaction rate as there no more enzyme (E) available for reacting with substrate (S). Here, the rate of reaction becomes dependent on the ES complex and the reaction becomes a unimolecular reaction with an order of zero. Though the enzymatic mechanism for the unimolecular reaction $ES -> [k_{cat}]E + P$ can be quite complex, there is typically one rate-determining enzymatic step that allows this reaction to be modelled as a single catalytic step with an apparent unimolecular rate constant k_{cat}. If the reaction path proceeds over one or several intermediates, k_{cat} will be a function of several elementary rate constants, whereas in the simplest case of a single elementary reaction (e.g. no intermediates) it will be identical to the elementary unimolecular rate constant k_2. The apparent unimolecular rate constant k_{cat} is also called *turnover number* and denotes the maximum number of enzymatic reactions catalysed per second.

The Michaelis–Menten equation describes how the (initial) reaction rate v_0 depends on the position of the substrate-binding equilibrium and the rate constant k_2.

$$v_0 = \frac{V_{\max}[S]}{K_M + [S]} \quad \textit{(Michaelis–Menten equation)}$$

with the constants

$$K_M \overset{\text{def}}{=} \frac{k_2 + k_{-1}}{k_1} \approx K_D$$

$$V_{\max} \overset{\text{def}}{=} k_{cat}[E]_{tot}$$

This Michaelis–Menten equation is the basis for most single-substrate enzyme kinetics. Two crucial assumptions underlie this equation (apart from the general assumption about the mechanism only involving no intermediate or product inhibition, and there is no allostericity or cooperativity). The first assumption is the so-called quasi-steady-state assumption (or pseudo-steady-state hypothesis), namely that the concentration of the substrate-bound enzyme (and hence also the unbound enzyme) changes much more slowly than those of the product and substrate and thus the change over time of the complex can be set to zero $d[ES]/dt = 0$. The second assumption is that the total enzyme concentration does not change over time, thus $[E]_{\text{tot}} = [E] + [ES] = \text{const.}$. A complete derivation can be found here.

The Michaelis constant K_M is experimentally defined as the concentration at which the rate of the enzyme reaction is half V_{max}, which can be verified by substituting $[S] = K_m$ into the Michaelis–Menten equation and can also be seen graphically. If the rate-determining enzymatic step is slow compared to substrate dissociation ($k_2 \ll k_{-1}$), the Michaelis constant K_M is roughly the dissociation constantK_D of the ES complex.

If $[S]$ is small compared to K_M then the term $[S]/(K_M + [S]) \approx [S]/K_M$ and also very little ES complex is formed, thus $[E]_0 \approx [E]$. Therefore, the rate of product formation is,

$$v_0 \approx \frac{k_{cat}}{K_M}[E][S] \qquad \text{if } [S] \ll K_M$$

Thus the product formation rate depends on the enzyme concentration as well as on the substrate concentration, the equation resembles a bimolecular reaction with a corresponding pseudo-second order rate constant k_2 / K_M. This constant is a measure of catalytic efficiency. The most efficient enzymes reach a k_2 / K_M in the range of 10^8 – 10^{10} M^{-1} s^{-1}. These enzymes are so efficient they effectively catalyse a reaction each time they encounter a substrate molecule and have thus reached an upper theoretical limit for efficiency (diffusion limit); and are sometimes referred to as kinetically perfect enzymes.

Direct use of the Michaelis–Menten Equation for Time Course Kinetic Analysis

The observed velocities predicted by the Michaelis–Menten equation can be used to directly model the time course disappearance of substrate and the production of product through incorporation of the Michaelis–Menten equation into the equation for first order chemical kinetics. This can only be achieved however if one recognises the problem associated with the use of Euler's number in the description of first order chemical kinetics. i.e. e^{-k} is a split constant that introduces a systematic error into calculations and can be rewritten as a single constant which represents the remaining substrate after each time period.

$$[S]=[S]_0(1-k)^t$$

$$[S]=[S]_0(1-v/[S]_0)^t$$

$$[S]=[S]_0(1-(V_{\max}[S]_0/(K_M+[S]_0)/[S]_0))^t$$

In 1983 Stuart Beal (and also independently Santiago Schnell and Claudio Mendoza in 1997) derived a closed form solution for the time course kinetics analysis of the Michaelis-Menten mechanism. The solution, known as the Schnell-Mendoza equation, has the form:

$$\frac{[S]}{K_M}=W\left[F(t)\right]$$

where W[] is the Lambert-W function. and where F(t) is,

$$F(t)=\frac{[S]_0}{K_M}\exp\left(\frac{[S]_0}{K_M}-\frac{V_{\max}}{K_M}t\right)$$

This equation is encompassed by the equation below, obtained by Berberan-Santos (MATCH Commun. Math. Comput. Chem. 63 (2010) 283), which is also valid when the initial substrate concentration is close to that of enzyme,

$$\frac{[S]}{K_M}=W\left[F(t)\right]-\frac{V_{\max}}{k_{cat}K_M}\frac{W\left[F(t)\right]}{1+W\left[F(t)\right]}$$

where W[] is again the Lambert-W function.

Linear Plots of the Michaelis–Menten Equation

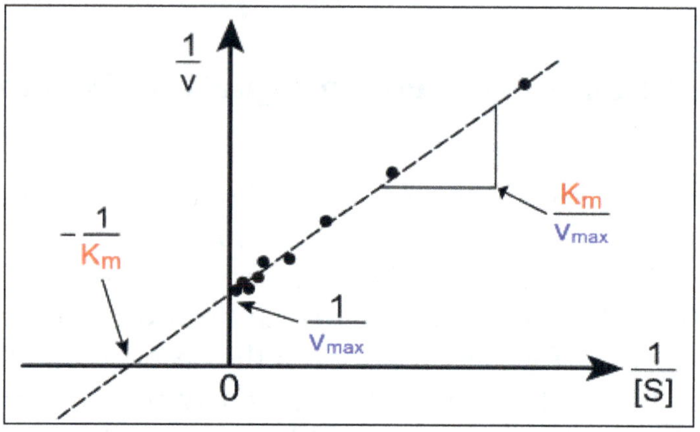

Lineweaver–Burk or double-reciprocal plot of kinetic data, showing the significance of the axis intercepts and gradient.

The plot of v versus [S] above is not linear; although initially linear at low [S], it bends over to saturate at high [S]. Before the modern era of nonlinear curve-fitting on computers, this nonlinearity could make it difficult to estimate K_M and V_{max} accurately. Therefore, several researchers developed linearisations of the Michaelis–Menten equation, such as the Lineweaver–Burk plot, the Eadie–Hofstee diagram and the Hanes–Woolf plot. All of these linear representations can be useful for visualising data, but none should be used to determine kinetic parameters, as computer software is readily available that allows for more accurate determination by nonlinear regression methods.

The Lineweaver–Burk plot or double reciprocal plot is a common way of illustrating kinetic data. This is produced by taking the reciprocal of both sides of the Michaelis–Menten equation. As shown on the right, this is a linear form of the Michaelis–Menten equation and produces a straight line with the equation $y = mx + c$ with a y-intercept equivalent to $1/V_{max}$ and an x-intercept of the graph representing $-1/K_M$.

$$\frac{1}{v}=\frac{K_M}{V_{max}[S]}+\frac{1}{V_{max}}$$

Naturally, no experimental values can be taken at negative 1/[S]; the lower limiting value 1/[S] = 0 (the y-intercept) corresponds to an infinite substrate concentration, where $1/v=1/V_{max}$ as shown at the right; thus, the x-intercept is an extrapolation of the experimental data taken at positive concentrations. More generally, the Lineweaver–Burk plot skews the importance of measurements taken at low substrate concentrations and, thus, can yield inaccurate estimates of V_{max} and K_M. A more accurate linear plotting method is the Eadie-Hofstee plot. In this case, v is plotted against v/[S]. In the third common linear representation, the Hanes-Woolf plot, [S]/v is plotted against [S]. In general, data normalisation can help diminish the amount of experimental work and can increase the reliability of the output, and is suitable for both graphical and numerical analysis.

Practical Significance of Kinetic Constants

The study of enzyme kinetics is important for two basic reasons. Firstly, it helps explain how enzymes work, and secondly, it helps predict how enzymes behave in living organisms. The kinetic constants defined above, K_M and V_{max}, are critical to attempts to understand how enzymes work together to control metabolism.

Making these predictions is not trivial, even for simple systems. For example, oxaloacetate is formed by malate dehydrogenase within the mitochondrion. Oxaloacetate can then be consumed by citrate synthase, phosphoenolpyruvate carboxykinase or aspartate aminotransferase, feeding into the citric acid cycle, gluconeogenesis or aspartic acid biosynthesis, respectively. Being able to predict how much oxaloacetate goes into which pathway requires knowledge of the concentration of oxaloacetate as well as the concentration and kinetics of each of these enzymes. This aim of predicting the behaviour of metabolic pathways reaches its most complex expression in the synthesis of huge amounts of kinetic and gene expression data into mathematical models of entire organisms. Alternatively, one useful simplification of the metabolic modelling problem is to ignore the underlying enzyme kinetics and only rely on information about the reaction network's stoichiometry, a technique called flux balance analysis.

Michaelis–Menten Kinetics with Intermediate

One could also consider the less simple case

$$\mathrm{E}+\mathrm{S} \underset{k_{-1}}{\overset{k_1}{\rightleftharpoons}} \mathrm{ES} \overset{k_2}{\rightarrow} \mathrm{EI} \overset{k_3}{\rightarrow} \mathrm{E}+\mathrm{P}$$

where a complex with the enzyme and an intermediate exists and the intermediate is converted into product in a second step. In this case we have a very similar equation

$$v_0 = k_{cat} \frac{[S][E]_0}{K_M^{'} + [S]}$$

but the constants are different

$$K_M^{'} \overset{\text{def}}{=} \frac{k_3}{k_2+k_3} K_M = \frac{k_3}{k_2+k_3} \cdot \frac{k_2+k_{-1}}{k_1}$$

$$k_{cat} \overset{\text{def}}{=} \frac{k_3 k_2}{k_2+k_3}$$

We see that for the limiting case $k_3 \gg k_2$, thus when the last step from $EI -> E+P$ is much faster than the previous step, we get again the original equation. Mathematically we have then $K_M^{'} \approx K_M$ and $k_{cat} \approx k_2$.

Multi-substrate Reactions

Multi-substrate reactions follow complex rate equations that describe how the substrates bind and in what sequence. The analysis of these reactions is much simpler if the concentration of substrate A is kept constant and substrate B varied. Under these conditions, the enzyme behaves just like a single-substrate enzyme and a plot of v by [S] gives apparent K_M and V_{max} constants for substrate B. If a set of these measurements is performed at different fixed concentrations of A, these data can be used to work out what the mechanism of the reaction is. For an enzyme that takes two substrates A and B and turns them into two products P and Q, there are two types of mechanism: ternary complex and ping–pong.

Ternary-complex Mechanisms

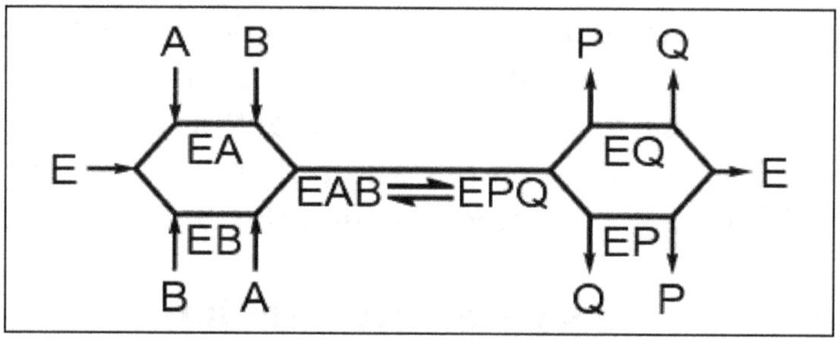

Random-order ternary-complex mechanism for an enzyme reaction. The reaction path is shown as a line and enzyme intermediates containing substrates A and B or products P and Q are written below the line.

In these enzymes, both substrates bind to the enzyme at the same time to produce an EAB ternary complex. The order of binding can either be random (in a random mechanism) or substrates have to bind in a particular sequence (in an ordered mechanism). When a set of v by [S] curves (fixed A, varying B) from an enzyme with a ternary-complex mechanism are plotted in a Lineweaver–Burk plot, the set of lines produced will intersect.

Enzymes with ternary-complex mechanisms include glutathione *S*-transferase,dihydrofolate reductase and DNA polymerase. The following links show short animations of the ternary-complex mechanisms of the enzymes dihydrofolate reductase[β] and DNA polymerase[γ].

Ping–pong Mechanisms

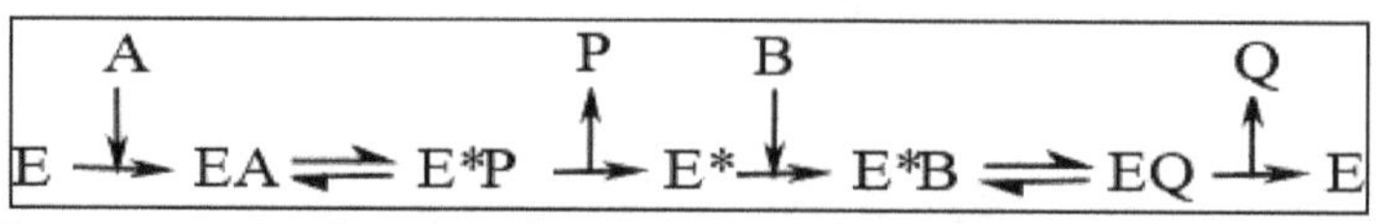

Ping–pong mechanism for an enzyme reaction. Intermediates contain substrates A and B or products P and Q.

As shown on the right, enzymes with a ping-pong mechanism can exist in two states, E and a chemically modified form of the enzyme E*; this modified enzyme is known as an intermediate. In such mechanisms, substrate A binds, changes the enzyme to E* by, for example, transferring a chemical group to the active site, and is then released. Only after the first substrate is released can substrate B bind and react with the modified enzyme, regenerating the unmodified E form. When a set of v by [S] curves (fixed A, varying B) from an enzyme with a ping–pong mechanism are plotted in a Lineweaver–Burk plot, a set of parallel lines will be produced. This is called a secondary plot.

Enzymes with ping–pong mechanisms include some oxidoreductases such as thioredoxin peroxidase,transferases such as acylneuraminate cytidylyltransferase and serine proteases such as trypsin and chymotrypsin. Serine proteases are a very common and diverse family of enzymes, including digestive enzymes (trypsin, chymotrypsin, and elastase), several enzymes of the blood clotting cascade and many others. In these serine proteases, the E* intermediate is an acyl-enzyme species formed by the attack of an active site serine residue on a peptide bond in a protein substrate. A short animation showing the mechanism of chymotrypsin is linked here.[8]

Non-Michaelis–Menten Kinetics

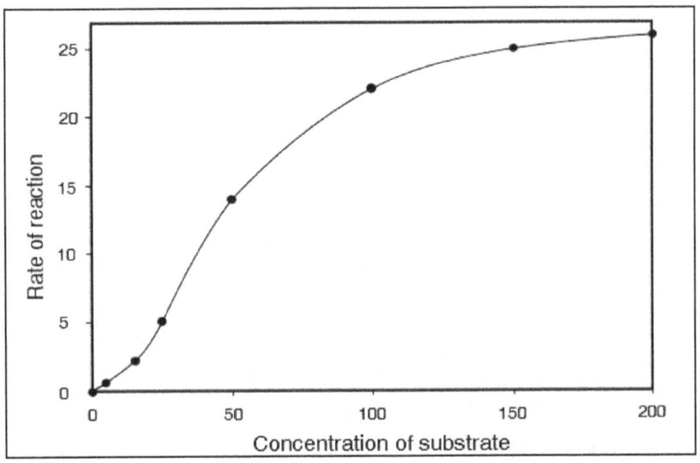

Saturation curve for an enzyme reaction showing sigmoid kinetics.

Some enzymes produce a sigmoid$v$ by [S] plot, which often indicates cooperative binding of substrate to the active site. This means that the binding of one substrate molecule affects the binding of subsequent substrate molecules. This behavior is most common in multimeric enzymes with several interacting active sites. Here, the mechanism of cooperation is similar to that of hemoglobin, with binding of substrate to one active site altering the affinity of the other active sites for substrate molecules. Positive cooperativity occurs when binding of the first substrate molecule *increases* the affinity of the other active sites for substrate. Negative cooperativity occurs when binding of the first substrate *decreases* the affinity of the enzyme for other substrate molecules.

Allosteric enzymes include mammalian tyrosyl tRNA-synthetase, which shows negative cooperativity, and bacterial aspartate transcarbamoylase and phosphofructokinase, which show positive cooperativity.

Cooperativity is surprisingly common and can help regulate the responses of enzymes to changes in the concentrations of their substrates. Positive cooperativity makes enzymes much more sensitive to [S] and their activities can show large changes over a narrow range of substrate concentration. Conversely, negative cooperativity makes enzymes insensitive to small changes in [S].

The Hill equation (biochemistry) is often used to describe the degree of cooperativity quantitatively in non-Michaelis–Menten kinetics. The derived Hill coefficient n measures how much the binding of substrate to one active site affects the binding of substrate to the other active sites. A Hill coefficient of <1 indicates negative cooperativity and a coefficient of >1 indicates positive cooperativity.

Pre-steady-state Kinetics

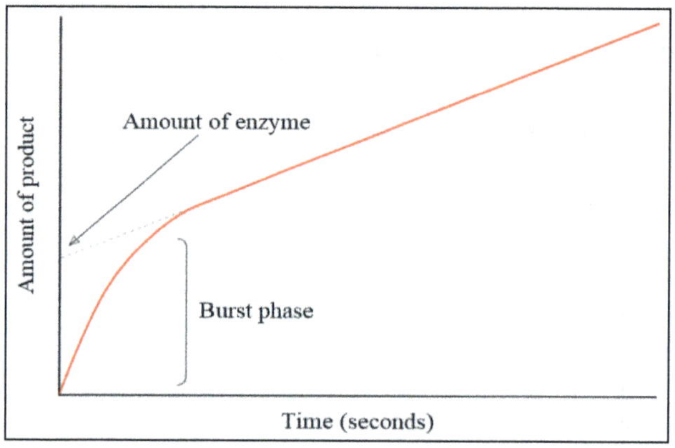

Pre-steady state progress curve, showing the burst phase of an enzyme reaction.

In the first moment after an enzyme is mixed with substrate, no product has been formed and no intermediates exist. The study of the next few milliseconds of the reaction is called pre-steady-state kinetics. Pre-steady-state kinetics is therefore concerned with the formation and consumption of enzyme–substrate intermediates (such as ES or E*) until their steady-state concentrations are reached.

This approach was first applied to the hydrolysis reaction catalysed by chymotrypsin. Often, the detection of an intermediate is a vital piece of evidence in investigations of what mechanism an enzyme follows. For example, in the ping–pong mechanisms that are shown above, rapid kinetic measurements can follow the release of product P and measure the formation of the modified enzyme intermediate E*. In the case of chymotrypsin, this intermediate is formed by an attack on the substrate by the nucleophilic serine in the active site and the formation of the acyl-enzyme intermediate.

In the figure to the right, the enzyme produces E* rapidly in the first few seconds of the reaction. The rate then slows as steady state is reached. This rapid burst phase of the reaction measures a single turnover of the enzyme. Consequently, the amount of product released in this burst, shown as the intercept on the y-axis of the graph, also gives the amount of functional enzyme which is present in the assay.

Chemical Mechanism

An important goal of measuring enzyme kinetics is to determine the chemical mechanism of an enzyme reaction, i.e., the sequence of chemical steps that transform substrate into product. The kinetic approaches discussed above will show at what rates intermediates are formed and inter-converted, but they cannot identify exactly what these intermediates are.

Kinetic measurements taken under various solution conditions or on slightly modified enzymes or substrates often shed light on this chemical mechanism, as they reveal the rate-determining step or intermediates in the reaction. For example, the breaking of a covalent bond to a hydrogenatom is a common rate-determining step. Which of the possible hydrogen transfers is rate determining can be shown by measuring the kinetic effects of substituting each hydrogen by deuterium, its stable isotope. The rate will change when the critical hydrogen is replaced, due to a primary kinetic isotope effect, which occurs because bonds to deuterium are harder to break than bonds to hydrogen. It is also possible to measure similar effects with other isotope substitutions, such as $^{13}C/^{12}C$ and $^{18}O/^{16}O$, but these effects are more subtle.

Isotopes can also be used to reveal the fate of various parts of the substrate molecules in the final products. For example, it is sometimes difficult to discern the origin of an oxygen atom in the final product; since it may have come from water or from part of the substrate. This may be determined by systematically substituting oxygen's stable isotope ^{18}O into the various molecules that participate in the reaction and checking for the isotope in the product. The chemical mechanism can also be elucidated by examining the kinetics and isotope effects under different pH conditions, by altering the metal ions or other bound cofactors, by site-directed mutagenesis of conserved amino acid residues, or by studying the behaviour of the enzyme in the presence of analogues of the substrate(s).

Enzyme Inhibition and Activation

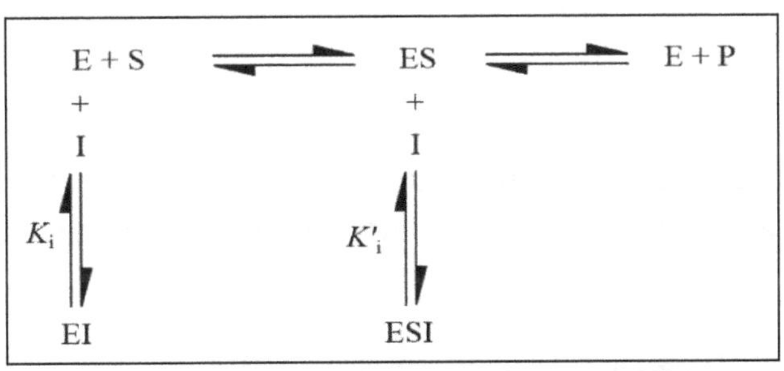

Kinetic scheme for reversible enzyme inhibitors.

Enzyme inhibitors are molecules that reduce or abolish enzyme activity, while enzyme activators are molecules that increase the catalytic rate of enzymes. These interactions can be either *reversible* (i.e., removal of the inhibitor restores enzyme activity) or *irreversible* (i.e., the inhibitor permanently inactivates the enzyme).

Reversible Inhibitors

Traditionally reversible enzyme inhibitors have been classified as competitive, uncompetitive, or non-competitive, according to their effects on K_m and V_{max}. These different effects result from the inhibitor binding to the enzyme E, to the enzyme–substrate complex ES, or to both, respectively. The division of these classes arises from a problem in their derivation and results in the need to use two different binding constants for one binding event. The binding of an inhibitor and its effect on the enzymatic activity are two distinctly different things, another problem the traditional equations fail to acknowledge. In noncompetitive inhibition the binding of the inhibitor results in 100% inhibition of the enzyme only, and fails to consider the possibility of anything in between. The common form of the inhibitory term also obscures the relationship between the inhibitor binding to the enzyme and its relationship to any other binding term be it the Michaelis–Menten equation or a dose response curve associated with ligand receptor binding. To demonstrate the relationship the following rearrangement can be made:

$$\frac{V_{\max}}{1+\frac{[I]}{K_i}}=\frac{V_{\max}}{\frac{[I]+K_i}{K_i}}$$

Adding zero to the bottom ([I]-[I])

$$\frac{V_{\max}}{\frac{[I]+K_i}{[I]+K_i-[I]}}$$

Dividing by $[I]+K_i$

$$\frac{V_{\max}}{\frac{1}{1-\frac{[I]}{[I]+K_i}}}=V_{\max}-V_{\max}\frac{[I]}{[I]+K_i}$$

This notation demonstrates that similar to the Michaelis–Menten equation, where the rate of reaction depends on the percent of the enzyme population interacting with substrate.

fraction of the enzyme population bound by substrate,

$$\frac{[S]}{[S]+K_m}$$

fraction of the enzyme population bound by inhibitor,

$$\frac{[I]}{[I]+K_i}$$

the effect of the inhibitor is a result of the percent of the enzyme population interacting with inhibitor. The only problem with this equation in its present form is that it assumes absolute inhibition of the enzyme with inhibitor binding, when in fact there can be a wide range of effects anywhere from 100% inhibition of substrate turn over to just >0%. To account for this the equation can be easily modified to allow for different degrees of inhibition by including a delta V_{max} term.

$$V_{max} - \Delta V_{max} \frac{[I]}{[I]+K_i}$$

or

$$V_{max1} - (V_{max1} - V_{max2}) \frac{[I]}{[I]+K_i}$$

This term can then define the residual enzymatic activity present when the inhibitor is interacting with individual enzymes in the population. However the inclusion of this term has the added value of allowing for the possibility of activation if the secondary V_{max} term turns out to be higher than the initial term. To account for the possibly of activation as well the notation can then be rewritten replacing the inhibitor "I" with a modifier term denoted here as "X".

$$V_{max1} - (V_{max1} - V_{max2}) \frac{[X]}{[X]+K_x}$$

While this terminology results in a simplified way of dealing with kinetic effects relating to the maximum velocity of the Michaelis–Menten equation, it highlights potential problems with the term used to describe effects relating to the K_m. The K_m relating to the affinity of the enzyme for the substrate should in most cases relate to potential changes in the binding site of the enzyme which would directly result from enzyme inhibitor interactions. As such a term similar to the one proposed above to modulate V_{max} should be appropriate in most situations:

$$K_{m1} - (K_{m1} - K_{m2}) \frac{[X]}{[X]+K_x}$$

Irreversible Inhibitors

Enzyme inhibitors can also irreversibly inactivate enzymes, usually by covalently modifying active site residues. These reactions, which may be called suicide substrates, follow exponential decay functions and are usually saturable. Below saturation, they follow first order kinetics with respect to inhibitor.

Mechanisms of Catalysis

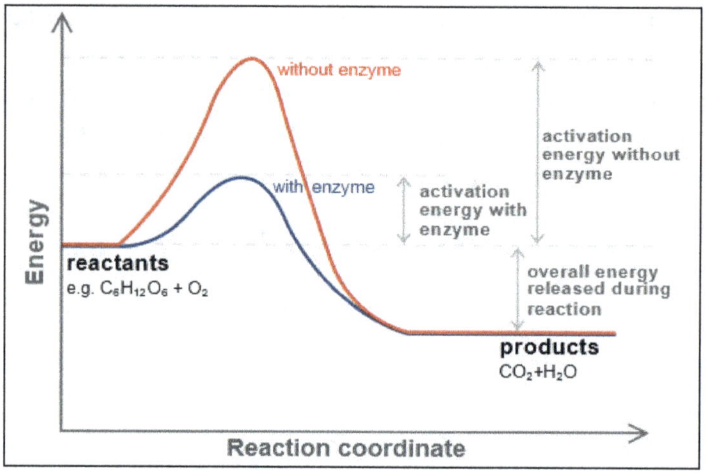

The energy variation as a function of reaction coordinate shows the stabilisation of the transition state by an enzyme.

The favoured model for the enzyme–substrate interaction is the induced fit model. This model proposes that the initial interaction between enzyme and substrate is relatively weak, but that these weak interactions rapidly induce conformational changes in the enzyme that strengthen binding. These conformational changes also bring catalytic residues in the active site close to the chemical bonds in the substrate that will be altered in the reaction. Conformational changes can be measured using circular dichroism or dual polarisation interferometry. After binding takes place, one or more mechanisms of catalysis lower the energy of the reaction's transition state by providing an alternative chemical pathway for the reaction. Mechanisms of catalysis include catalysis by bond strain; by proximity and orientation; by active-site proton donors or acceptors; covalent catalysis and quantum tunnelling.

Enzyme kinetics cannot prove which modes of catalysis are used by an enzyme. However, some kinetic data can suggest possibilities to be examined by other techniques. For example, a ping–pong mechanism with burst-phase pre-steady-state kinetics would suggest covalent catalysis might be important in this enzyme's mechanism. Alternatively, the observation of a strong pH effect on V_{max} but not K_m might indicate that a residue in the active site needs to be in a particular ionisation state for catalysis to occur.

History

In 1902 Victor Henri proposed a quantitative theory of enzyme kinetics, but at the time the experimental significance of the hydrogen ion concentration was not yet recognized. After Peter Lauritz Sørensen had defined the logarithmic pH-scale and introduced the concept of buffering in 1909 the German chemist Leonor Michaelis and his Canadian postdoc Maud Leonora Menten repeated Henri's experiments and confirmed his equation, which is now generally referred to as Michaelis-Menten kinetics (sometimes also *Henri-Michaelis-Menten kinetics*). Their work was further developed by G. E. Briggs and J. B. S. Haldane, who derived kinetic equations that are still widely considered today a starting point in modeling enzymatic activity.

The major contribution of the Henri-Michaelis-Menten approach was to think of enzyme reactions in two stages. In the first, the substrate binds reversibly to the enzyme, forming the enzyme-substrate complex. This is sometimes called the Michaelis complex. The enzyme then catalyzes the chemical step in the reaction and releases the product. The kinetics of many enzymes is adequately described by the simple Michaelis-Menten model, but all enzymes have internal motions that are not accounted for in the model and can have significant contributions to the overall reaction kinetics. This can be modeled by introducing several Michaelis-Menten pathways that are connected with fluctuating rates, which is a mathematical extension of the basic Michaelis Menten mechanism.

Software

ENZO

ENZO (Enzyme Kinetics) is a graphical interface tool for building kinetic models of enzyme catalyzed reactions. ENZO automatically generates the corresponding differential equations from a stipulated enzyme reaction scheme. These differential equations are processed by a numerical solver and a regression algorithm which fits the coefficients of differential equations to experimentally observed time course curves. ENZO allows rapid evaluation of rival reaction schemes and can be used for routine tests in enzyme kinetics.

Molecular Motor

Molecular motors are biological molecular machines that are the essential agents of movement in living organisms. In general terms, a motor is a device that consumes energy in one form and converts it into motion or mechanical work; for example, many protein-based molecular motors harness the chemical free energy released by the hydrolysis of ATP in order to perform mechanical work. In terms of energetic efficiency, this type of motor can be superior to currently available man-made motors. One important difference between molecular motors and macroscopic motors is that molecular motors operate in the thermal bath, an environment in which the fluctuations due to thermal noise are significant.

Examples

Some examples of biologically important molecular motors:

- Cytoskeletal motors
  - Myosins are responsible for muscle contraction, intracellular cargo transport, and producing cellular tension.
  - Kinesin moves cargo inside cells away from the nucleus along microtubules.
  - Dynein produces the axonemal beating of cilia and flagella and also transports cargo along microtubules towards the cell nucleus.
- Polymerisation motors
  - Actin polymerization generates forces and can be used for propulsion. ATP is used.
  - Microtubule polymerization using GTP.
  - Dynamin is responsible for the separation of clathrin buds from the plasma membrane. GTP is used.
- Rotary motors:
  - F_0F_1-ATP synthase family of proteins convert the chemical energy in ATP to the electrochemical potential energy of a proton gradient across a membrane or the other way around. The catalysis of the chemical reaction and the movement of protons are coupled to each other via the mechanical rotation of parts of the complex. This is involved in ATP synthesis in the mitochondria and chloroplasts as well as in pumping of protons across the vacuolar membrane.
  - The bacterial flagellum responsible for the swimming and tumbling of *E. coli* and other bacteria acts as a rigid propeller that is powered by a rotary motor. This motor is driven by the flow of protons across a membrane, possibly using a similar mechanism to that found in the F_0 motor in ATP synthase.
- Nucleic acid motors:
  - RNA polymerase transcribes RNA from a DNA template.
  - DNA polymerase turns single-stranded DNA into double-stranded DNA.
  - Helicases separate double strands of nucleic acids prior to transcription or replication. ATP is used.

- Topoisomerases reduce supercoiling of DNA in the cell. ATP is used.
- RSC and SWI/SNF complexes remodel chromatin in eukaryotic cells. ATP is used.
- SMC proteins responsible for chromosome condensation in eukaryotic cells.
- Viral DNA packaging motors inject viral genomic DNA into capsids as part of their replication cycle, packing it very tightly. Several models have been put forward to explain how the protein generates the force required to drive the DNA into the capsid; for a review . An alternative proposal is that, in contrast with all other biological motors, the force is not generated directly by the protein, but by the DNA itself. In this model, ATP hydrolysis is used to drive protein conformational changes that alternatively dehydrate and rehydrate the DNA, cyclically driving it from B-DNA to A-DNA and back again. A-DNA is 23% shorter than B-DNA, and the DNA shrink/expand cycle is coupled to a protein-DNA grip/release cycle to generate the forward motion that propels DNA into the capsid.

- Synthetic molecular motors have been created by chemists that yield rotation, possibly generating torque.

Theoretical Considerations

Because the motor events are stochastic, molecular motors are often modeled with the Fokker-Planck equation or with Monte Carlo methods. These theoretical models are especially useful when treating the molecular motor as a Brownian motor.

Experimental Observation

In experimental biophysics, the activity of molecular motors is observed with many different experimental approaches, among them:

- Fluorescent methods: fluorescence resonance energy transfer (FRET), fluorescence correlation spectroscopy (FCS), total internal reflection fluorescence (TIRF).
- Magnetic tweezers can also be useful for analysis of motors that operate on long pieces of DNA.
- Neutron spin echo spectroscopy can be used to observe motion on nanosecond timescales.
- Optical tweezers are well-suited for studying molecular motors because of their low spring constants.

- Scattering techniques: single particle tracking based on dark field microscopy or interferometric scattering microscopy (iSCAT).
- Single-molecule electrophysiology can be used to measure the dynamics of individual ion channels.

Many more techniques are also used. As new technologies and methods are developed, it is expected that knowledge of naturally occurring molecular motors will be helpful in constructing synthetic nanoscale motors.

Non-biological

Recently, chemists and those involved in nanotechnology have begun to explore the possibility of creating molecular motors *de novo*. These synthetic molecular motors currently suffer many limitations that confine their use to the research laboratory. However, many of these limitations may be overcome as our understanding of chemistry and physics at the nanoscale increases. Systems like the nanocars, while not technically motors, are illustrative of recent efforts towards synthetic nanoscale motors.

Myoglobin

Myoglobin (symbol Mb or MB) is an iron- and oxygen-binding protein found in the muscle tissue of vertebrates in general and in almost all mammals. It is related to hemoglobin, which is the iron- and oxygen-binding protein in blood, specifically in the red blood cells. In humans, myoglobin is only found in the bloodstream after muscle injury. It is an abnormal finding, and can be diagnostically relevant when found in blood.

Myoglobin is the primary oxygen-carrying pigment of muscle tissues. High concentrations of myoglobin in muscle cells allow organisms to hold their breath for a longer period of time. Diving mammals such as whales and seals have muscles with particularly high abundance of myoglobin. Myoglobin is found in Type I muscle, Type II A and Type II B, but most texts consider myoglobin not to be found in smooth muscle.

Myoglobin was the first protein to have its three-dimensional structure revealed by X-ray crystallography. This achievement was reported in 1958 by John Kendrew and associates. For this discovery, John Kendrew shared the 1962 Nobel Prize in chemistry with Max Perutz. Despite being one of the most studied proteins in biology, its physiological function is not yet conclusively established: mice genetically engineered to lack myoglobin are viable, but showed a 30% reduction in volume of blood being pumped by the heart during a contraction. They adapted to this deficiency through natural reactions to inadequate oxygen supply (hypoxia) and a widening of blood vessels (vasodilation). In humans myoglobin is encoded by the *MB*gene.

As hemoglobin can take the forms of oxyhemoglobin (HbO_2), carboxyhemoglobin (HbCO), and methemoglobin (met-Hb), so can myoglobin take the forms of oxymyoglobin (MbO_2), carboxymyoglobin (MbCO), and metmyoglobin (met-Mb).

Differences from Hemoglobin

Myoglobin is similar to hemoglobin in that it is involved in the transportation of oxygen to cells. There are many distinct differences that set the protein apart from hemoglobin. For one the protein has only one binding site for oxygen on the one heme group on the protein. While myoglobin can only hold one oxygen, the affinity for that oxygen is very high compared to hemoglobin. This is probably because hemoglobin, transporting 4 oxygens to the tissues and muscles where myoglobin is mostly present. The myoglobin takes the oxygen from the hemoglobin (due to the Bohr Effect) and takes that oxygen to muscle cells for use in metabolic processes.

Meat Color

Myoglobin contains hemes, pigments responsible for the color of red meat. The color that meat takes is partly determined by the degree of oxidation of the myoglobin. In fresh meat the iron atom is the ferrous (+2) oxidation state bound to an oxygen molecule (O_2). Meat cooked well done is brown because the iron atom is now in the ferric (+3) oxidation state, having lost an electron. If meat has been exposed to nitrites, it will remain pink because the iron atom is bound to NO, nitric oxide (true of, e.g., corned beef or cured hams). Grilled meats can also take on a pink "smoke ring" that comes from the iron binding to a molecule of carbon monoxide. Raw meat packed in a carbon monoxide atmosphere also shows this same pink "smoke ring" due to the same principles. Notably, the surface of this raw meat also displays the pink color, which is usually associated in consumers' minds with fresh meat. This artificially induced pink color can persist, reportedly up to one year. Hormel and Cargill are both reported to use this meat-packing process, and meat treated this way has been in the consumer market since 2003.

Role in Disease

Myoglobin is released from damaged muscle tissue (rhabdomyolysis), which has very high concentrations of myoglobin. The released myoglobin is filtered by the kidneys but is toxic to the renal tubular epithelium and so may cause acute kidney injury. It is not the myoglobin itself that is toxic (it is a protoxin) but the ferrihemate portion that is dissociated from myoglobin in acidic environments (e.g., acidic urine, lysosomes).

Myoglobin is a sensitive marker for muscle injury, making it a potential marker for heart attack in patients with chest pain. However, elevated myoglobin has low specificity for acute myocardial infarction (AMI) and thus CK-MB, cardiac Troponin, ECG, and clinical signs should be taken into account to make the diagnosis.

Structure and Bonding

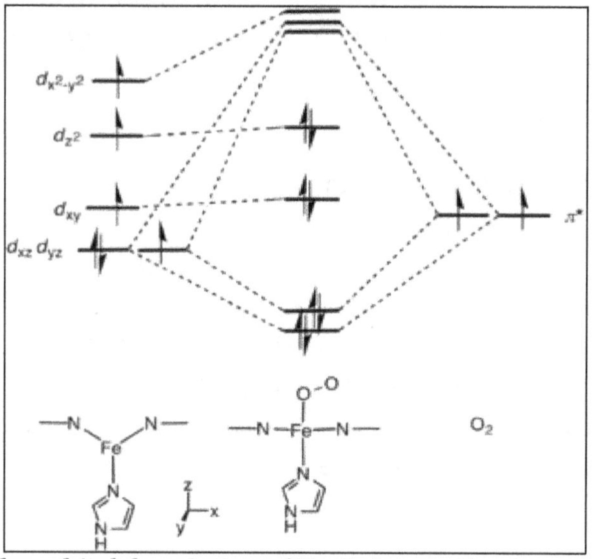

Molecular orbital description of Fe-O_2 interaction in myoglobin.

Myoglobin belongs to the globin superfamily of proteins, and as with other globins, consists of eight alpha helices connected by loops. Human globin contains 154 amino acids.

Myoglobin contains a porphyrin ring with an iron at its center. A *proximal*histidine group (His-94) is attached directly to iron, and a *distal* histidine group (His-65) hovers near the opposite face. The distal imidazole is not bonded to the iron but is available to interact with the substrate O_2. This interaction encourages the binding of O_2, but not carbon monoxide (CO), which still binds about 240× more strongly than O_2.

The binding of O_2 causes substantial structural change at the Fe center, which shrinks in radius and moves into the center of N4 pocket. O_2-binding induces "spin-pairing": the five-coordinate ferrous deoxy form is high spin and the six coordinate oxy form is low spin and diamagnetic.

Synthetic Analogues

Many models of myoglobin have been synthesized as part of a broad interest in transition metal dioxygen complexes. A well known example is the *picket fence porphyrin*, which consists of a ferrous complex of a sterically bulky derivative of tetraphenylporphyrin. In the presence of an imidazole ligand, this ferrous complex reversibly binds O_2. The O_2 substrate adopts a bent geometry, occupying the sixth position of the iron center. A key property of this model is the slow formation of the μ-oxo dimer, which is an inactive diferric state. In nature, such deactivation pathways are suppressed by protein matrix that prevents close approach of the Fe-porphyrin assemblies.

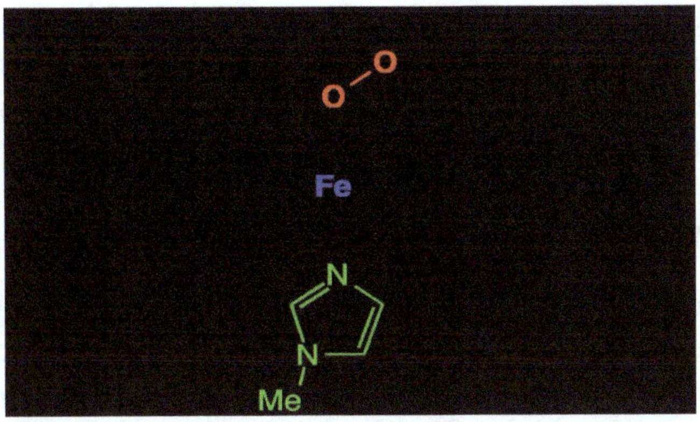

A picket-fence porphyrin complex of Fe, with axial coordination sites occupied by methylimidazole (green) and dioxygen. The R groups flank the O_2-binding site.

Hemoglobin

Hemoglobin also spelledhaemoglobin (United Kingdom spelling) and abbreviated Hb or Hgb, is the iron-containing oxygen-transport metalloprotein in the red blood cells of all vertebrates (with the exception of the fish family Channichthyidae) as well as the tissues of some invertebrates. Hemoglobin in the blood carries oxygen from the respiratory organs (lungs or gills) to the rest of the body (i.e. the tissues). There it releases the oxygen to permit aerobic respiration to provide energy to power the functions of the organism in the process called metabolism.

In mammals, the protein makes up about 96% of the red blood cells' dry content (by weight), and around 35% of the total content (including water). Hemoglobin has an oxygen-binding capacity of 1.34 mL O_2 per gram, which increases the total blood oxygen capacity seventy-fold compared to dissolved oxygen in blood. The mammalian hemoglobin molecule can bind (carry) up to four oxygen molecules.

Hemoglobin is involved in the transport of other gases: It carries some of the body's respiratory carbon dioxide (about 20–25% of the total) as carbaminohemoglobin, in which CO_2 is bound to the globin protein. The molecule also carries the important regulatory molecule nitric oxide bound to a globin protein thiol group, releasing it at the same time as oxygen.

Hemoglobin is also found outside red blood cells and their progenitor lines. Other cells that contain hemoglobin include the A9 dopaminergic neurons in the substantia nigra, macrophages, alveolar cells, and mesangial cells in the kidney. In these tissues, hemoglobin has a non-oxygen-carrying function as an antioxidant and a regulator of iron metabolism.

Hemoglobin and hemoglobin-like molecules are also found in many invertebrates, fungi, and plants. In these organisms, hemoglobins may carry oxygen, or they may act to transport and regulate other small molecules and ions such as carbon dioxide, nitric oxide, hydrogen sulfide and sulfide. A variant of the molecule, called leghemoglobin, is used to scavenge oxygen away from anaerobic systems, such as the nitrogen-fixing nodules of leguminous plants, before the oxygen can poison (deactivate) the system.

Research History

In 1825 J.F. Engelhard discovered that the ratio of Fe to protein is identical in the hemoglobins of several species. From the known atomic mass of iron he calculated the molecular mass of hemoglobin to $n \times 16000$ (n = number of iron atoms per hemoglobin, now known to be 4), the first determination of a protein's molecular mass. This "hasty conclusion" drew a lot of ridicule at the time from scientists who could not believe that any molecule could be that big. Gilbert Smithson Adair confirmed Engelhard's results in 1925 by measuring the osmotic pressure of hemoglobin solutions.

The oxygen-carrying protein hemoglobin was discovered by Hünefeld in 1840. In 1851, German physiologist Otto Funke published a series of articles in which he described growing hemoglobin crystals by successively diluting red blood cells with a solvent such as pure water, alcohol or ether, followed by slow evaporation of the solvent from the resulting protein solution. Hemoglobin's reversible oxygenation was described a few years later by Felix Hoppe-Seyler.

In 1959, Max Perutz determined the molecular structure of myoglobin (similar to hemoglobin) by X-ray crystallography. This work resulted in his sharing with John Kendrew the 1962 Nobel Prize in Chemistry.

The role of hemoglobin in the blood was elucidated by French physiologistClaude Bernard. The name *hemoglobin* is derived from the words *heme* and *globin*, reflecting the fact that each subunit of hemoglobin is a globular protein with an embedded heme group. Each heme group contains one iron atom, that can bind one oxygen molecule through ion-induced dipole forces. The most common type of hemoglobin in mammals contains four such subunits.

Genetics

Hemoglobin consists of protein subunits (the "globin" molecules), and these proteins, in turn, are folded chains of a large number of different amino acids called polypeptides. The amino acid sequence of any polypeptide created by a cell is in turn determined by the stretches of DNA called genes. In all proteins, it is the amino acid sequence that determines the protein's chemical properties and function.

There is more than one hemoglobin gene: in humans, hemoglobin A (the main form of hemoglobin present) is coded for by the genes, *HBA1*, *HBA2*, and *HBB*. The amino acid sequences of the globin proteins in hemoglobins usually differ between species. These differences grow with evolutionary distance between species. For example, the most common hemoglobin sequences in humans and chimpanzees are nearly identical, differing by only one amino acid in both the alpha and the beta globin protein chains. These differences grow larger between less closely related species.

Even within a species, different variants of hemoglobin always exist, although one sequence is usually a "most common" one in each species. Mutations in the genes for the hemoglobin protein in a species result in hemoglobin variants. Many of these mutant forms of hemoglobin cause no disease. Some of these mutant forms of hemoglobin, however, cause a group of hereditary diseases termed the *hemoglobinopathies*. The best known hemoglobinopathy is sickle-cell disease, which was the first human disease whose mechanism was understood at the molecular level. A (mostly) separate set of diseases called thalassemias involves underproduction of normal and sometimes abnormal hemoglobins, through problems and mutations in globin gene regulation. All these diseases produce anemia.

Variations in hemoglobin amino acid sequences, as with other proteins, may be adaptive. For example, hemoglobin has been found to adapt in different ways to high altitudes. Organisms living at high elevations experience lower partial pressures of oxygen compared to those at sea level. This presents a challenge to the organisms that inhabit such environments because hemoglobin, which normally binds oxygen at high partial pressures of oxygen, must be able to bind oxygen when it is present at a lower pressure. Different organisms have adapted to such a challenge. For example, recent studies have suggested genetic variants in deer mice that help explain how deer mice that live in the mountains are able to survive in the thin air that accompanies high altitudes. A researcher from the University of Nebraska-Lincoln found mutations in four different genes that can account for differences between deer mice that live in lowland prairies versus the mountains. After examining wild mice captured from both highlands and lowlands, it was found that: the genes of the two breeds are "virtually identical–except for those that govern the oxygen-carrying capacity of their hemoglobin". "The genetic difference enables highland mice to make more efficient use of their oxygen", since less is available at higher altitudes, such as those in the mountains.Mammoth hemoglobin featured mutations that allowed for oxygen delivery at lower temperatures, thus enabling mammoths to migrate to higher latitudes during the Pleistocene. This was also found in hummingbirds that inhabit the Andes. Hummingbirds already expend a lot of energy and thus have high oxygen demands and yet Andean hummingbirds have been found to thrive in high altitudes. Non-synonymous mutations in the hemoglobin gene of multiple species living at high elevations (*Oreotrochilus, A. castelnaudii, C. violifer, P. gigas,* and *A. viridicuada*) have caused the protein to have less of an affinity for inositol hexaphosphate (IHP), a molecule found in birds that has a similar role as 2,3-BPG in humans; this results in the ability to bind oxygen in lower partial pressures.

It should be noted, however, that birds' unique circulatory lungs also promote efficient use of oxygen at low partial pressures of O_2. These two adaptations reinforce each other and account for birds' remarkable high-altitude performance.

Hemoglobin adaptation extends to humans, as well. Studies have found that a small number of native Tibetan women have a genotype which codes for hemoglobin to be more highly saturated with oxygen. Natural selection seems to be the main force working on this gene because mortality rate of offspring is significantly lower for women with higher hemoglobin-oxygen affinity when compared to the mortality rate of offspring from women with low hemoglobin-oxygen affinity. While the exact genotype and mechanism by which this occurs is not yet clear, selection is acting on these women's ability to bind oxygen in low partial pressures which overall allows them to better sustain crucial metabolic processes.

Synthesis

Hemoglobin (Hb) is synthesized in a complex series of steps. The heme part is synthesized in a series of steps in the mitochondria and the cytosol of immature red blood cells, while the globin protein parts are synthesized by ribosomes in the cytosol. Production of Hb continues in the cell throughout its early development from the proerythroblast to the reticulocyte in the bone marrow. At this point, the nucleus is lost in mammalian red blood cells, but not in birds and many other species. Even after the loss of the nucleus in mammals, residual ribosomal RNA allows further synthesis of Hb until the reticulocyte loses its RNA soon after entering the vasculature (this hemoglobin-synthetic RNA in fact gives the reticulocyte its reticulated appearance and name).

Structure

Hemoglobin has a quaternary structure characteristic of many multi-subunit globular proteins. Most of the amino acids in hemoglobin form alpha helices, connected by short non-helical segments. Hydrogen bonds stabilize the helical sections inside this protein, causing attractions within the molecule, folding each polypeptide chain into a specific shape. Hemoglobin's quaternary structure comes from its four subunits in roughly a tetrahedral arrangement.

In most vertebrates, the hemoglobin molecule is an assembly of four globular protein subunits. Each subunit is composed of a protein chain tightly associated with a non-protein prostheticheme group. Each protein chain arranges into a set of alpha-helix structural segments connected together in a globin fold arrangement, so called because this arrangement is the same folding motif used in other heme/globin proteins such as myoglobin. This folding pattern contains a pocket that strongly binds the heme group.

A heme group consists of an iron (Fe) ion (charged atom) held in a heterocyclic ring, known as a porphyrin. This porphyrin ring consists of four pyrrole molecules cyclically linked together (by methine bridges) with the iron ion bound in the center. The iron ion, which is the site of oxygen binding, coordinates with the four nitrogen atoms in the center of the ring, which all lie in one plane. The iron is bound strongly (covalently) to the globular protein via the N atoms of the imidazole ring of F8 histidine residue (also known as the proximal histidine) below the porphyrin ring. A sixth position can reversibly bind oxygen by a coordinate covalent bond, completing the octahedral group of six ligands. Oxygen binds in an "end-on bent" geometry where one oxygen atom binds to Fe and the other protrudes at an angle. When oxygen is not bound, a very weakly bonded water molecule fills the site, forming a distorted octahedron.

Even though carbon dioxide is carried by hemoglobin, it does not compete with oxygen for the iron-binding positions but is bound to the protein chains of the structure.

The iron ion may be either in the Fe^{2+} or in the Fe^{3+} state, but ferrihemoglobin (methemoglobin) (Fe^{3+}) cannot bind oxygen. In binding, oxygen temporarily and reversibly oxidizes (Fe^{2+}) to (Fe^{3+}) while oxygen temporarily turns into the superoxide ion, thus iron must exist in the +2 oxidation state to bind oxygen. If superoxide ion associated to Fe^{3+} is protonated, the hemoglobin iron will remain oxidized and incapable of binding oxygen. In such cases, the enzyme methemoglobin reductase will be able to eventually reactivate methemoglobin by reducing the iron center.

In adult humans, the most common hemoglobin type is a tetramer (which contains four subunit proteins) called *hemoglobin A*, consisting of two α and two β subunits non-covalently bound, each made of 141 and 146 amino acid residues, respectively. This is denoted as $\alpha_2\beta_2$. The subunits are structurally similar and about the same size. Each subunit has a molecular weight of about 16,000 daltons, for a total molecular weight of the tetramer of about 64,000 daltons (64,458 g/mol). Thus, 1 g/dL = 0.1551 mmol/L. Hemoglobin A is the most intensively studied of the hemoglobin molecules.

In human infants, the hemoglobin molecule is made up of 2 α chains and 2 γ chains. The gamma chains are gradually replaced by β chains as the infant grows.

The four polypeptide chains are bound to each other by salt bridges, hydrogen bonds, and the hydrophobic effect.

Oxygen Saturation

In general, hemoglobin can be saturated with oxygen molecules (oxyhemoglobin), or desaturated with oxygen molecules (deoxyhemoglobin).

Oxyhemoglobin

Oxyhemoglobin is formed during physiological respiration when oxygen binds to the heme component of the protein hemoglobin in red blood cells. This process occurs in the pulmonary capillaries adjacent to the alveoli of the lungs. The oxygen then travels through the blood stream to be dropped off at cells where it is utilized as a terminal electron acceptor in the production of ATP by the process of oxidative phosphorylation. It does not, however, help to counteract a decrease in blood pH. Ventilation, or breathing, may reverse this condition by removal of carbon dioxide, thus causing a shift up in pH.

Hemoglobin exists in two forms, a *taut (tense) form* (T) and a *relaxed form* (R). Various factors such as low pH, high CO_2 and high 2,3 BPG at the level of the tissues favor the taut form, which has low oxygen affinity and releases oxygen in the tissues. Conversely, a high pH, low CO_2, or low 2,3 BPG favors the relaxed form, which can better bind oxygen. The partial pressure of the system also affects O_2 affinity where, at high partial pressures of oxygen (such as those present in the alveoli), the relaxed (high affinity, R) state is favoured. Inversely, at low partial pressures (such as those present in respiring tissues), the (low affinity, T) tense state is favoured. Additionally, the binding of oxygen to the iron(II) heme pulls the iron into the plane of the porphyrin ring, causing a slight conformational shift. The shift encourages oxygen to bind to the three remaining heme units within hemoglobin (thus, oxygen binding is cooperative).

Deoxygenated Hemoglobin

Deoxygenated hemoglobin is the form of hemoglobin without the bound oxygen. The absorption spectra of oxyhemoglobin and deoxyhemoglobin differ. The oxyhemoglobin has significantly lower absorption of the 660 nm wavelength than deoxyhemoglobin, while at 940 nm its absorption is slightly higher. This difference is used for the measurement of the amount of oxygen in a patient's blood by an instrument called a pulse oximeter. This difference also accounts for the presentation of cyanosis, the blue to purplish color that tissues develop during hypoxia.

Evolution of Vertebrate Hemoglobin

Scientists agree that the event that separated myoglobin from hemoglobin occurred after lampreys diverged from jawed vertebrates. This separation of myoglobin and hemoglobin allowed for the different functions of the two molecules to arise and develop: myoglobin has more to do with oxygen storage while hemoglobin is tasked with oxygen transport. The α- and β-like globin genes encode the individual subunits of the protein. The predecessors of these genes arose through another duplication event also after the gnathosome common ancestor derived from jawless fish, approximately 450–500 million years ago. The development of α and β genes created the potential for hemoglobin to be composed of multiple subunits, a physical composition central to hemoglobin's ability to transport oxygen. Having multiple subunits contributes to hemoglobin's ability to bind oxygen cooperatively as well as be regulated allosterically. Subsequently, the α gene also underwent a duplication event to form the *HBA1* and *HBA2* genes. These further duplications and divergences have created a diverse range of α- and β-like globin genes that are regulated so that certain forms occur at different stages of development.

Iron's Oxidation State in Oxyhemoglobin

Assigning oxygenated hemoglobin's oxidation state is difficult because oxyhemoglobin (Hb-O_2), by experimental measurement, is diamagnetic (no net unpaired electrons), yet the lowest-energy (ground-state) electron configurations in both oxygen and iron are paramagnetic (suggesting at least one unpaired electron in the complex). The lowest-energy form of oxygen, and the lowest energy forms of the relevant oxidation states of iron, are these:

- Triplet oxygen, the lowest-energy molecular oxygen species, has two unpaired electrons in antibonding π^* molecular orbitals.
- Iron(II) tends to exist in a high-spin $3d^6$ configuration with four unpaired electrons.
- Iron(III) ($3d^5$) has an odd number of electrons, and thus must have one or more unpaired electrons, in any energy state.

All of these structures are paramagnetic (have unpaired electrons), not diamagnetic. Thus, a non-intuitive (e.g., a higher-energy for at least one species) distribution of electrons in the combination of iron and oxygen must exist, in order to explain the observed diamagnetism and no unpaired electrons.

The two logical possibilities to produce diamagnetic (no net spin) Hb-O_2 are:

1. Low-spin Fe^{2+} binds to singlet oxygen. Both low-spin iron and singlet oxygen are diamagnetic. However, the singlet form of oxygen is the higher-energy form of the molecule.
2. Low-spin Fe^{3+} binds to $O_2^{\bullet-}$ (the superoxide ion) and the two unpaired electrons couple antiferromagnetically, giving observed diamagnetic properties. Here, the iron has been oxidized (has lost one electron), and the oxygen has been reduced (has gained one electron).

Another possible model in which low-spin Fe^{4+} binds to peroxide, O_2^{2-}, can be ruled out by itself, because the iron is paramagnetic (although the peroxide ion is diamagnetic). Here, the iron has been oxidized by two electrons, and the oxygen reduced by two electrons.

Direct Experimental Data:

- X-ray photoelectron spectroscopy suggests iron has an oxidation state of approximately 3.2.
- Infrared vibrational frequencies of the O-O bond suggests a bond length fitting with superoxide (a bond order of about 1.6, with superoxide being 1.5).
- X-ray Absorption Near Edge Structures at the iron K-edge. The energy shift of 5 eV between deoxyhemoglobin and oxyhemoglobin, as for all the methemoglobin species, strongly suggests an actual local charge closer to Fe^{3+} than Fe^{2+}.

Thus, the nearest formal oxidation state of iron in Hb-O_2 is the +3 state, with oxygen in the −1 state (as superoxide $.O_2^-$). The diamagnetism in this configuration arises from the single unpaired electron on superoxide aligning antiferromagnetically with the single unpaired electron on iron (in a low-spin d^5 state), to give no net spin to the entire configuration, in accordance with diamagnetic oxyhemoglobin from experiment.

The second choice of the logical possibilities above for diamagnetic oxyhemoglobin being found correct by experiment, is not surprising: singlet oxygen (possibility #1) is an unrealistically high energy state. Model 3 leads to unfavorable separation of charge (and does not agree with the magnetic data), although it could make a minor contribution as a resonance form. Iron's shift to a higher oxidation state in Hb-O_2 decreases the atom's size, and allows it into the plane of the porphyrin ring, pulling on the coordinated histidine residue and initiating the allosteric changes seen in the globulins.

Early postulates by bio-inorganic chemists claimed that possibility #1 (above) was correct and that iron should exist in oxidation state II. This conclusion seemed likely, since the iron oxidation state III as methemoglobin, when *not* accompanied by superoxide $.O_2^-$ to "hold" the oxidation electron, was known to render hemoglobin incapable of binding normal triplet O_2 as it occurs in the air. It was thus assumed that iron remained as Fe(II) when oxygen gas was bound in the lungs. The iron chemistry in this previous classical model was elegant, but the required presence of the diamagnetic, high-energy, singlet oxygen molecule was never explained. It was classically argued that the binding of an oxygen molecule placed high-spin iron(II) in an octahedral field of strong-field ligands; this change in field would increase the crystal field splitting energy, causing iron's electrons to pair into the low-spin configuration, which would be diamagnetic in Fe(II). This forced low-spin pairing is indeed thought to happen in iron when oxygen binds, but is not enough to explain iron's change in size. Extraction of an additional electron from iron by oxygen is required to explain both iron's smaller size and observed increased oxidation state, and oxygen's weaker bond.

The assignment of a whole-number oxidation state is a formalism, as the covalent bonds are not required to have perfect bond orders involving whole electron transfer. Thus, all three models for paramagnetic Hb-O_2 may contribute to some small degree (by resonance) to the actual electronic configuration of Hb-O_2. However, the model of iron in Hb-O_2 being Fe(III) is more correct than the classical idea that it remains Fe(II).

Cooperativity

A schematic visual model of oxygen-binding process, showing all four monomers and hemes, and protein chains only as diagramatic coils, to facilitate visualization into the molecule. Oxygen is not shown in this model, but, for each of the iron atoms, it binds to the iron (red sphere) in the flat heme. For example, in the upper-left of the four hemes shown, oxygen binds at the left of the iron atom shown in the upper-left of diagram. This causes the iron atom to move backward into the heme that holds it (the iron moves upward as it binds oxygen, in this illustration), tugging the histidine residue (modeled as a red pentagon on the right of the iron) closer, as it does. This, in turn, pulls on the protein chain holding the histidine.

When oxygen binds to the iron complex, it causes the iron atom to move back toward the center of the plane of the porphyrin ring. At the same time, the imidazole side-chain of the histidine residue interacting at the other pole of the iron is pulled toward the porphyrin ring. This interaction forces the plane of the ring sideways toward the outside of the tetramer, and also induces a strain in the protein helix containing the histidine as it moves nearer to the iron atom. This strain is transmitted to the remaining three monomers in the tetramer, where it induces a similar conformational change in the other heme sites such that binding of oxygen to these sites becomes easier.

In the tetrameric form of normal adult hemoglobin, the binding of oxygen is, thus, a cooperative process. The binding affinity of hemoglobin for oxygen is increased by the oxygen saturation of the molecule, with the first oxygens bound influencing the shape of the binding sites for the next oxygens, in a way favorable for binding. This positive cooperative binding is achieved through steric conformational changes of the hemoglobin protein complex as discussed above; i.e., when one subunit protein in hemoglobin becomes oxygenated, a conformational or structural change in the whole complex is initiated, causing the other subunits to gain an increased affinity for oxygen. As a consequence, the oxygen binding curve of hemoglobin is sigmoidal, or *S*-shaped, as opposed to the normal hyperbolic curve associated with noncooperative binding.

The dynamic mechanism of the cooperativity in hemoglobin and its relation with the low-frequency resonance has been discussed.

Binding for Ligands other than Oxygen

Besides the oxygen ligand, which binds to hemoglobin in a cooperative manner, hemoglobin ligands also include competitive inhibitors such as carbon monoxide (CO) and allosteric ligands such as carbon dioxide (CO_2) and nitric oxide (NO). The carbon dioxide is bound to amino groups of the globin proteins as carbaminohemoglobin, and is thought to account for about 10% of carbon dioxide transport in mammals. Nitric oxide is bound to specific thiol groups in the globin protein to form an S-nitrosothiol, which dissociates into free nitric oxide and thiol again, as the hemoglobin releases oxygen from its heme site. This nitric oxide transport to peripheral tissues is hypothesized to assist oxygen transport in tissues, by releasing vasodilatory nitric oxide to tissues in which oxygen levels are low.

Competitive

The binding of oxygen is affected by molecules such as carbon monoxide (CO) (for example, from tobacco smoking, car exhaust, and incomplete combustion in furnaces).

CO competes with oxygen at the heme binding site. Hemoglobin binding affinity for CO is 250 times greater than its affinity for oxygen, meaning that small amounts of CO dramatically reduce hemoglobin's ability to transport oxygen. Since carbon monoxide is a colorless, odorless and tasteless gas, and poses a potentially fatal threat, detectors have become commercially available to warn of dangerous levels in residences. When hemoglobin combines with CO, it forms a very bright red compound called carboxyhemoglobin, which may cause the skin of CO poisoning victims to appear pink in death, instead of white or blue. When inspired air contains CO levels as low as 0.02%, headache and nausea occur; if the CO concentration is increased to 0.1%, unconsciousness will follow. In heavy smokers, up to 20% of the oxygen-active sites can be blocked by CO.

In similar fashion, hemoglobin also has competitive binding affinity for cyanide (CN^-), sulfur monoxide (SO), and sulfide (S^{2-}), including hydrogen sulfide (H_2S). All of these bind to iron in heme without changing its oxidation state, but they nevertheless inhibit oxygen-binding, causing grave toxicity.

The iron atom in the heme group must initially be in the ferrous (Fe^{2+}) oxidation state to support oxygen and other gases' binding and transport (it temporarily switches to ferric during the time oxygen is bound, as explained above). Initial oxidation to the ferric (Fe^{3+}) state without oxygen converts hemoglobin into "hem**i**globin" or methemoglobin (pronounced "MET-hemoglobin"), which cannot bind oxygen. Hemoglobin in normal red blood cells is protected by a reduction system to keep this from happening. Nitric oxide is capable of converting a small fraction of hemoglobin to methemoglobin in red blood cells. The latter reaction is a remnant activity of the more ancient nitric oxide dioxygenase function of globins.

Allosteric

Carbon *di*oxide occupies a different binding site on the hemoglobin. Carbon dioxide is more readily dissolved in deoxygenated blood, facilitating its removal from the body after the oxygen has been released to tissues undergoing metabolism. This increased affinity for carbon dioxide by the venous blood is known as the Haldane effect. Through the enzyme carbonic anhydrase, carbon dioxide reacts with water to give carbonic acid, which decomposes into bicarbonate and protons:

$$CO_2 + H_2O \rightarrow H_2CO_3 \rightarrow HCO_3^- + H^+$$

The sigmoidal shape of hemoglobin's oxygen-dissociation curve results from cooperative binding of oxygen to hemoglobin.

Hence, blood with high carbon dioxide levels is also lower in pH (more acidic). Hemoglobin can bind protons and carbon dioxide, which causes a conformational change in the protein and facilitates the release of oxygen. Protons bind at various places on the protein, while carbon dioxide binds at the α-amino group. Carbon dioxide binds to hemoglobin and forms carbaminohemoglobin. This decrease in hemoglobin's affinity for oxygen by the binding of carbon dioxide and acid is known as the Bohr effect (shifts the O_2-saturation curve to the *right*). Conversely, when the carbon dioxide levels in the blood decrease (i.e., in the lung capillaries), carbon dioxide and protons are released from hemoglobin, increasing the oxygen affinity of the protein. A reduction in the total binding capacity of hemoglobin to oxygen (i.e. shifting the curve down, not just to the right) due to reduced pH is called the root effect. This is seen in bony fish.

It is necessary for hemoglobin to release the oxygen that it binds; if not, there is no point in binding it. The sigmoidal curve of hemoglobin makes it efficient in binding (taking up O_2 in lungs), and efficient in unloading (unloading O_2 in tissues).

In people acclimated to high altitudes, the concentration of 2,3-Bisphosphoglycerate (2,3-BPG) in the blood is increased, which allows these individuals to deliver a larger amount of oxygen to tissues under conditions of lower oxygen tension. This phenomenon, where molecule Y affects the binding of molecule X to a transport molecule Z, is called a *heterotropic*allosteric effect. Hemoglobin in organisms at high altitudes has also adapted such that it has less of an affinity for 2,3-BPG and so the protein will be shifted more towards its R state. In its R state, hemoglobin will bind oxygen more readily, thus allowing organisms to perform the necessary metabolic processes when oxygen is present at low partial pressures.

Animals other than humans use different molecules to bind to hemoglobin and change its O_2 affinity under unfavorable conditions. Fish use both ATP and GTP. These bind to a phosphate "pocket" on the fish hemoglobin molecule, which stabilizes the tense state and therefore decreases oxygen affinity. GTP reduces hemoglobin oxygen affinity much more than ATP, which is thought to be due to an extra hydrogen bond formed that further stabilizes the tense state. Under hypoxic conditions, the concentration of both ATP and GTP is reduced in fish red blood cells to increase oxygen affinity.

A variant hemoglobin, called fetal hemoglobin (HbF, $\alpha_2\gamma_2$), is found in the developing fetus, and binds oxygen with greater affinity than adult hemoglobin. This means that the oxygen binding curve for fetal hemoglobin is left-shifted (i.e., a higher percentage of hemoglobin has oxygen bound to it at lower oxygen tension), in comparison to that of adult hemoglobin. As a result, fetal blood in the placenta is able to take oxygen from maternal blood.

Hemoglobin also carries nitric oxide (NO) in the globin part of the molecule. This improves oxygen delivery in the periphery and contributes to the control of respiration. NO binds reversibly to a specific cysteine residue in globin; the binding depends on the state (R or T) of the hemoglobin. The resulting S-nitrosylated hemoglobin influences various NO-related activities such as the control of vascular resistance, blood pressure and respiration. NO is not released in the cytoplasm of erythrocytes but transported by an anion exchanger called AE1 out of them.

Types in Humans

Hemoglobin variants are a part of the normal embryonic and fetal development. They may also be pathologic mutant forms of hemoglobin in a population, caused by variations in genetics. Some well-known hemoglobin variants, such as sickle-cell anemia, are responsible for diseases and are considered hemoglobinopathies. Other variants cause no detectable pathology, and are thus considered non-pathological variants.

In the embryo:

- Gower 1 ($\zeta_2\varepsilon_2$)
- Gower 2 ($\alpha_2\varepsilon_2$) (PDB: 1A9W)
- Hemoglobin Portland I ($\zeta_2\gamma_2$)
- Hemoglobin Portland II ($\zeta_2\beta_2$).

In the fetus:

- Hemoglobin F ($\alpha_2\gamma_2$) (PDB: 1FDH).

After birth:

- Hemoglobin A ($\alpha_2\beta_2$) (PDB: 1BZ0) – The most common with a normal amount over 95%.
- Hemoglobin A_2 ($\alpha_2\delta_2$) – δ chain synthesis begins late in the third trimester and, in adults, it has a normal range of 1.5–3.5%.
- Hemoglobin F ($\alpha_2\gamma_2$) – In adults Hemoglobin F is restricted to a limited population of red cells called F-cells. However, the level of Hb F can be elevated in persons with sickle-cell disease and beta-thalassemia.

Gene expression of hemoglobin before and after birth. Also identifies the types of cells and organs in which the gene expression (data on *Wood W.G.*, (1976). Br. Med. Bull. 32, 282.)

Variant forms that cause disease:

- Hemoglobin D-Punjab – ($\alpha_2\beta^D{}_2$) – A variant form of hemoglobin.
- Hemoglobin H (β_4) – A variant form of hemoglobin, formed by a tetramer of β chains, which may be present in variants of α thalassemia.
- Hemoglobin Barts (γ_4) – A variant form of hemoglobin, formed by a tetramer of γ chains, which may be present in variants of α thalassemia.
- Hemoglobin S ($\alpha_2\beta^S{}_2$) – A variant form of hemoglobin found in people with sick-

le cell disease. There is a variation in the β-chain gene, causing a change in the properties of hemoglobin, which results in sickling of red blood cells.

- Hemoglobin C ($\alpha_2\beta^C_2$) – Another variant due to a variation in the β-chain gene. This variant causes a mild chronic hemolytic anemia.
- Hemoglobin E ($\alpha_2\beta^E_2$) – Another variant due to a variation in the β-chain gene. This variant causes a mild chronic hemolytic anemia.
- Hemoglobin AS – A heterozygous form causing sickle cell trait with one adult gene and one sickle cell disease gene.
- Hemoglobin SC disease – A compound heterozygous form with one sickle gene and another encoding Hemoglobin C.

Degradation in Vertebrate Animals

When red cells reach the end of their life due to aging or defects, they are removed from the circulation by the phagocytic activity of macrophages in the spleen or the liver or hemolyze within the circulation. Free hemoglobin is then cleared from the circulation via the hemoglobin transporter CD163, which is exclusively expressed on monocytes or macrophages. Within these cells the hemoglobin molecule is broken up, and the iron gets recycled. This process also produces one molecule of carbon monoxide for every molecule of heme degraded. Heme degradation is one of the few natural sources of carbon monoxide in the human body, and is responsible for the normal blood levels of carbon monoxide even in people breathing pure air. The other major final product of heme degradation is bilirubin. Increased levels of this chemical are detected in the blood if red cells are being destroyed more rapidly than usual. Improperly degraded hemoglobin protein or hemoglobin that has been released from the blood cells too rapidly can clog small blood vessels, especially the delicate blood filtering vessels of the kidneys, causing kidney damage. Iron is removed from heme and salvaged for later use, it is stored as hemosiderin or ferritin in tissues and transported in plasma by beta globulins as transferrins. When the porphyrin ring is broken up, the fragments are normally secreted as a yellow pigment called bilirubin, which is secreted into the intestines as bile. Intestines metabolise bilirubin into urobilinogen. Urobilinogen leaves the body in faeces, in a pigment called stercobilin. Globulin is metabolised into amino acids that are then released into circulation.

Role in Disease

Hemoglobin deficiency can be caused either by a decreased amount of hemoglobin molecules, as in anemia, or by decreased ability of each molecule to bind oxygen at the same partial pressure of oxygen. Hemoglobinopathies (genetic defects resulting in abnormal structure of the hemoglobin molecule) may cause both. In any case, hemoglobin deficiency decreases blood oxygen-carrying capacity. Hemoglobin deficiency is, in general, strictly distinguished from hypoxemia, defined as decreased partial pressure of oxygen in blood, although both are causes of hypoxia (insufficient oxygen supply to tissues).

Other common causes of low hemoglobin include loss of blood, nutritional deficiency, bone marrow problems, chemotherapy, kidney failure, or abnormal hemoglobin (such as that of sickle-cell disease).

The ability of each hemoglobin molecule to carry oxygen is normally modified by altered blood pH or CO_2, causing an altered oxygen–hemoglobin dissociation curve. However, it can also be pathologically altered in, e.g., carbon monoxide poisoning.

Decrease of hemoglobin, with or without an absolute decrease of red blood cells, leads to symptoms of anemia. Anemia has many different causes, although iron deficiency and its resultant iron deficiency anemia are the most common causes in the Western world. As absence of iron decreases heme synthesis, red blood cells in iron deficiency anemia are *hypochromic* (lacking the red hemoglobin pigment) and *microcytic* (smaller than normal). Other anemias are rarer. In hemolysis (accelerated breakdown of red blood cells), associated jaundice is caused by the hemoglobin metabolite bilirubin, and the circulating hemoglobin can cause renal failure.

Some mutations in the globin chain are associated with the hemoglobinopathies, such as sickle-cell disease and thalassemia. Other mutations, as discussed at the beginning of the article, are benign and are referred to merely as hemoglobin variants.

There is a group of genetic disorders, known as the *porphyrias* that are characterized by errors in metabolic pathways of heme synthesis. King George III of the United Kingdom was probably the most famous porphyria sufferer.

To a small extent, hemoglobin A slowly combines with glucose at the terminal valine (an alpha aminoacid) of each β chain. The resulting molecule is often referred to as Hb A_{1c}. As the concentration of glucose in the blood increases, the percentage of Hb A that turns into Hb A_{1c} increases. In diabetics whose glucose usually runs high, the percent Hb A_{1c} also runs high. Because of the slow rate of Hb A combination with glucose, the Hb A_{1c} percentage is representative of glucose level in the blood averaged over a longer time (the half-life of red blood cells, which is typically 50–55 days).

Glycosylated hemoglobin is the form of hemoglobin to which glucose is bound. The binding of glucose to amino acids in the hemoglobin takes place spontaneously (without the help of an enzyme) in many proteins, and is not known to serve a useful purpose. However, the binding to hemoglobin does serve as a record for average blood glucose levels over the lifetime of red cells, which is approximately 120 days. The levels of glycosylated hemoglobin are therefore measured in order to monitor the long-term control of the chronic disease of type 2 diabetes mellitus (T2DM). Poor control of T2DM results in high levels of glycosylated hemoglobin in the red blood cells. The normal reference range is approximately 4–5.9 %. Though difficult to obtain, values less than 7% are recommended for people with T2DM. Levels greater than 9% are associated with poor control of the glycosylated hemoglobin, and levels greater than 12% are associated with very poor control. Diabetics who keep their glycosylated hemoglobin levels close to 7% have a much better chance of avoiding the complications that may accompany diabetes (than those whose levels are 8% or higher). In addition, increased glycosylation of hemoglobin increases its affinity for oxygen, therefore preventing its release at the tissue and inducing a level of hypoxia in extreme cases.

Elevated levels of hemoglobin are associated with increased numbers or sizes of red blood cells, called polycythemia. This elevation may be caused by congenital heart disease, cor pulmonale, pulmonary fibrosis, too much erythropoietin, or polycythemia vera. High hemoglobin levels may also be caused by exposure to high altitudes, smoking, dehydration (artificially by concentrating Hb), advanced lung disease and certain tumors.

A recent study done in Pondicherry, India, shows its importance in coronary artery disease.

Diagnostic Uses

Hemoglobin concentration measurement is among the most commonly performed blood tests, usually as part of a complete blood count. For example, it is typically tested before or after blood donation. Results are reported in g/L, g/dL or mol/L. 1 g/dL equals about 0.6206 mmol/L, although the latter units are not used as often due to uncertainty regarding the polymeric state of the molecule. This conversion factor, using the single globin unit molecular weight of 16,000 Da, is more common for hemoglobin concentration in blood. For MCHC (mean corpuscular hemoglobin concentration) the conversion factor 0.155, which uses the tetramer weight of 64,500 Da, is more common. Normal levels are:

- Men: 13.8 to 18.0 g/dL (138 to 180 g/L, or 8.56 to 11.17 mmol/L)
- Women: 12.1 to 15.1 g/dL (121 to 151 g/L, or 7.51 to 9.37 mmol/L)
- Children: 11 to 16 g/dL (111 to 160 g/L, or 6.83 to 9.93 mmol/L)
- Pregnant women: 11 to 14 g/dL (110 to 140 g/L, or 6.83 to 8.69 mmol/L) (9.5 to 15 usual value during pregnancy)

Normal values of hemoglobin in the 1st and 3rd trimesters of pregnant women must be at least 11 g/dL and at least 10.5 g/dL during the 2nd trimester.

Dehydration or hyperhydration can greatly influence measured hemoglobin levels. Albumin can indicate hydration status.

If the concentration is below normal, this is called anemia. Anemias are classified by the size of red blood cells, the cells that contain hemoglobin in vertebrates. The anemia is called “microcytic” if red cells are small, “macrocytic” if they are large, and “normocytic” otherwise.

Hematocrit, the proportion of blood volume occupied by red blood cells, is typically about three times the hemoglobin concentration measured in g/dL. For example, if the hemoglobin is measured at 17 g/dL, that compares with a hematocrit of 51%.

Laboratory hemoglobin test methods require a blood sample (arterial, venous, or capillary) and analysis on hematology analyzer and CO-oximeter. Additionally, a new noninvasive hemoglobin (SpHb) test method called Pulse CO-Oximetry is also available with comparable accuracy to invasive methods.

Concentrations of oxy- and deoxyhemoglobin can be measured continuously, regionally and noninvasively using NIRS. NIRS can be used both on the head as on muscles. This technique is often used for research in e.g. elite sports training, ergonomics, rehabilitation, patient monitoring, neonatal research, functional brain monitoring, brain computer interface, urology (bladder contraction), neurology (Neurovascular coupling) and more.

Long-term control of blood sugar concentration can be measured by the concentration of Hb A_{1c}. Measuring it directly would require many samples because blood sugar levels vary widely through the day. Hb A_{1c} is the product of the irreversible reaction of hemoglobin A with glucose. A higher glucose concentration results in more Hb A_{1c}. Because the reaction is slow, the Hb A_{1c} proportion represents glucose level in blood averaged over the half-life of red blood cells, is typically 50–55 days. An Hb A_{1c} proportion of 6.0% or less show good long-term glucose control, while values above 7.0% are elevated. This test is especially useful for diabetics.

The functional magnetic resonance imaging (fMRI) machine uses the signal from deoxyhemoglobin, which is sensitive to magnetic fields since it is paramagnetic. Combined measurement with NIRS shows good correlation with both the oxy- and deoxyhemoglobin signal compared to the BOLD signal.

Athletic Tracking and Self Tracking Uses

Hemoglobin can be tracked noninvasively, to build an individual data set tracking the hemoconcentration and hemodilution effects of daily activities for better understanding of sports performance and training. Athletes are often concerned about endurance and intensity of exercise. Using the scientific technique of absorption spectroscopy with eight wavelengths of light. This method is similar to a pulse oximeter, which consists of a small sensing device that clips to the finger. The sensor uses light-emitting diodes that emit red and infrared light through the tissue to a light detector, which then sends a signal to a processor to calculate the absorption of light by the hemoglobin protein.

Analogues in Non-vertebrate Organisms

A variety of oxygen-transport and -binding proteins exist in organisms throughout the animal and plant kingdoms. Organisms including bacteria, protozoans, and fungi all have hemoglobin-like proteins whose known and predicted roles include the reversible binding of gaseous ligands. Since many of these proteins contain globins and the heme moiety (iron in a flat porphyrin support), they are often called hemoglobins, even if their overall tertiary structure is very different from that of vertebrate hemoglobin. In particular, the distinction of "myoglobin" and hemoglobin in lower animals is often impossible, because some of these organisms do not contain muscles. Or, they may have a recognizable separate circulatory system but not one that deals with oxygen transport (for example, many insects and other arthropods). In all these groups, heme/globin-containing molecules (even monomeric globin ones) that deal with gas-binding are referred to as oxyhemoglobins. In addition to dealing with transport and sensing of oxygen, they may also deal with NO, CO_2, sulfide compounds, and even O_2 scavenging in environments that must be anaerobic. They may even deal with detoxification of chlorinated materials in a way analogous to heme-containing P450 enzymes and peroxidases.

The giant tube worm *Riftia pachyptila* showing red hemoglobin-containing plumes

The structure of hemoglobins varies across species. Hemoglobin occurs in all kingdoms of organisms, but not in all organisms. Primitive species such as bacteria, protozoa, algae, and plants often have single-globin hemoglobins. Many nematode worms, molluscs, and crustaceans contain very large multisubunit molecules, much larger than those in vertebrates. In particular, chimeric hemoglobins found in fungi and giant annelids may contain both globin and other types of proteins.

One of the most striking occurrences and uses of hemoglobin in organisms is in the giant tube worm (*Riftia pachyptila*, also called Vestimentifera), which can reach 2.4 meters length and populates ocean volcanic vents. Instead of a digestive tract, these worms contain a population of bacteria constituting half the organism's weight. The bacteria react with H_2S from the vent and O_2 from the water to produce energy to make food from H_2O and CO_2. The worms end with a deep-red fan-like structure ("plume"), which extends into the water and absorbs H_2S and O_2 for the bacteria, and CO_2 for use as synthetic raw material similar to photosynthetic plants. The structures are bright-red due to their containing several extraordinarily complex hemoglobins that have up to 144 globin chains, each including associated heme structures. These hemoglobins are remarkable for being able to carry oxygen in the presence of sulfide, and even to carry sulfide, without being completely "poisoned" or inhibited by it as hemoglobins in most other species are.

Other Oxygen-binding Proteins

Myoglobin

Found in the muscle tissue of many vertebrates, including humans, it gives muscle tissue a distinct red or dark gray color. It is very similar to hemoglobin in structure and sequence, but is not a tetramer; instead, it is a monomer that lacks cooperative binding. It is used to store oxygen rather than transport it.

Hemocyanin

The second most common oxygen-transporting protein found in nature, it is found in the blood of many arthropods and molluscs. Uses copper prosthetic groups instead of iron heme groups and is blue in color when oxygenated.

Hemerythrin

Some marine invertebrates and a few species of annelid use this iron-containing non-heme protein to carry oxygen in their blood. Appears pink/violet when oxygenated, clear when not.

Chlorocruorin

Found in many annelids, it is very similar to erythrocruorin, but the heme group is significantly different in structure. Appears green when deoxygenated and red when oxygenated.

Vanabins

Also known as *vanadium chromagens*, they are found in the blood of sea squirts. They were once hypothesized to use the rare metal vanadium as an oxygen binding prosthetic group. However, although they do contain vanadium by preference, they apparently bind little oxygen, and thus have some other function, which has not been elucidated (sea squirts also contain some hemoglobin). They may act as toxins.

Erythrocruorin

Found in many annelids, including earthworms, it is a giant free-floating blood protein containing many dozens—possibly hundreds—of iron- and heme-bearing protein subunits bound together into a single protein complex with a molecular mass greater than 3.5 million daltons.

Pinnaglobin

Only seen in the mollusc *Pinna nobilis*. Brown manganese-based porphyrin protein.

Leghemoglobin

In leguminous plants, such as alfalfa or soybeans, the nitrogen fixing bacteria in the roots are protected from oxygen by this iron heme containing oxygen-binding protein. The specific enzyme protected is nitrogenase, which is unable to reduce nitrogen gas in the presence of free oxygen.

Coboglobin

A synthetic cobalt-based porphyrin. Coboprotein would appear colorless when oxygenated, but yellow when in veins.

Presence in Nonerythroid Cells

Some nonerythroid cells (i.e., cells other than the red blood cell line) contain hemoglobin. In the brain, these include the A9 dopaminergic neurons in the substantia nigra, astrocytes in the cerebral cortex and hippocampus, and in all mature oligodendrocytes. It has been suggested that brain hemoglobin in these cells may enable the "storage of oxygen to provide a homeostatic mechanism in anoxic conditions, which is especially important for A9 DA neurons that have an elevated metabolism with a high requirement for energy production". It has been noted further that "A9 dopaminergic neurons may be at particular risk since in addition to their high mitochondrial activity they are under intense oxidative stress caused by the production of hydrogen peroxide via autoxidation and/or monoamine oxidase (MAO)-mediated deamination of dopamine and the subsequent reaction of accessible ferrous iron to generate highly toxic hydroxyl radicals". This may explain the risk of these cells for degeneration in Parkinson's disease. The hemoglobin-derived iron in these cells is not the cause of the post-mortem darkness of these cells (origin of the Latin name, substantia *nigra*), but rather is due to neuromelanin.

Outside the brain, hemoglobin has non-oxygen-carrying functions as an antioxidant and a regulator of iron metabolism in macrophages,alveolar cells, and mesangial cells in the kidney.

In History, Art and Music

Heart of Steel (Hemoglobin) (2005) by Julian Voss-Andreae. The images show the 5-foot (1.60 m) tall sculpture right after installation, after 10 days, and after several months of exposure to the elements.

Historically, an association between the color of blood and rust occurs in the association of the planet Mars, with the Roman god of war, since the planet is an orange-red, which reminded the ancients of blood. Although the color of the planet is due to iron compounds in combination with oxygen in the Martian soil, it is a common misconception that the iron in hemoglobin and its oxides gives blood its red color. The color is actually due to the porphyrinmoiety of hemoglobin to which the iron is bound, not the iron itself, although the ligation and redox state of the iron can influence the pi to pi* or n to pi* electronic transitions of the porphyrin and hence its optical characteristics.

Artist Julian Voss-Andreae created a sculpture called "Heart of Steel (Hemoglobin)" in 2005, based on the protein's backbone. The sculpture was made from glass and weathering steel. The intentional rusting of the initially shiny work of art mirrors hemoglobin's fundamental chemical reaction of oxygen binding to iron.

Montreal artist Nicolas Baier created Lustre (Hémoglobine), a sculpture in stainless steel that shows the structure of the hemoglobin molecule. It is displayed in the atrium of McGill University Health Centre's research centre in Montreal. The sculpture measures about 10 metres × 10 metres × 10 metres.

Pancreatic Ribonuclease

Pancreatic ribonucleases (EC3.1.27.5, *RNase, RNase I, RNase A, pancreatic RNase, ribonuclease I, endoribonuclease I, ribonucleic phosphatase, alkaline ribonuclease, ribonuclease, gene S glycoproteins, Ceratitis capitata alkaline ribonuclease, SLSG glycoproteins, gene S locus-specific glycoproteins, S-genotype-assocd. glycoproteins, ribonucleate 3'-pyrimidino-oligonucleotidohydrolase*) are pyrimidine-specific endonucleases found in high quantity in the pancreas of certain mammals and of some reptiles.

Specifically, the enzymes are involved in endonucleolytic cleavage of 3'-phosphomononucleotides and 3'-phosphooligonucleotides ending in C-P or U-P with 2',3'-cyclic phosphate intermediates. Ribonuclease can unwind the RNA helix by complexing with single-stranded RNA; the complex arises by an extended multi-site cation-anion interaction between lysine and arginine residues of the enzyme and phosphate groups of the nucleotides.

Notable Family Members

Bovine pancreatic ribonuclease is the best-studied member of the family and has served as a model system in work related to protein folding, disulfide bond formation, protein crystallography and spectroscopy, and protein dynamics.

Other proteins belonging to the pancreatic ribonuclease superfamily include: bovine seminal vesicle and brain ribonucleases; kidney non-secretory ribonucleases; liver-type ribonucleases;angiogenin, which induces vascularisation of normal and malignant tissues; eosinophil cationic protein, a cytotoxin and helminthotoxin with ribonuclease activity; and frog liver ribonuclease and frog sialic acid-binding lectin. The sequence of pancreatic ribonucleases contains four conserved disulfide bonds and three amino acid residues involved in the catalytic activity.

Human Genes

Humangenes encoding proteins containing this domain include:

- ANG,
- RNASE1, RNASE10, RNASE12, RNASE2, RNASE3, RNASE4, RNASE6, RNASE7, and RNASE8.

Cytotoxicity

Some members of the pancreatic ribonuclease family have cytotoxic effects. Mammalian cells are protected from these effects due to their extremely high affinity for ribonuclease inhibitor (RI), which protects cellular RNA from degradation by pancreatic ribonucleases. Pancreatic ribonucleases that are not inhibited by RI are approximately as toxic as alpha-sarcin, diphtheria toxin, or ricin.

Two pancreatic ribonucleases isolated from the oocytes of the Northern leopard frog - amphinase and ranpirnase - are not inhibited by RI and show differential cytotoxicity against tumor cells. Ranpirnase was studied in a Phase IIIclinical trial as a treatment candidate for mesothelioma, but the trial did not demonstrate statistical significance against primary endpoints.

Antibody

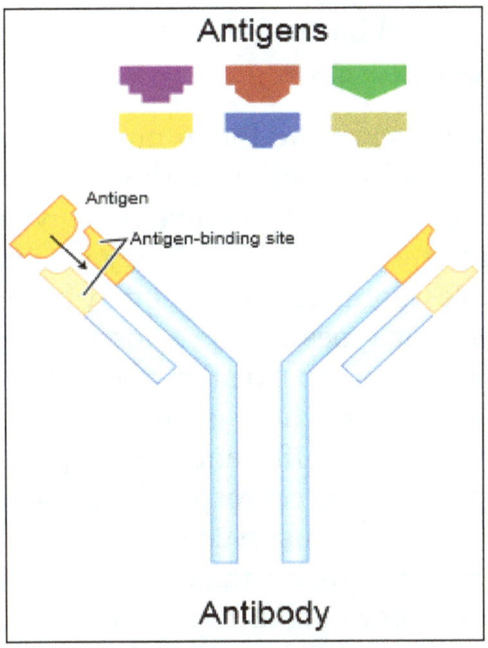

Each antibody binds to a specific antigen; an interaction similar to a lock and key.

An antibody (Ab), also known as an immunoglobulin (Ig), is a large, Y-shaped protein produced mainly by plasma cells that is used by the immune system to identify and neutralize pathogens such as bacteria and viruses. The antibody recognizes a unique molecule of the harmful agent, called an antigen, via the Fab's variable region. Each tip of the "Y" of an antibody contains a paratope (analogous to a lock) that is specific for one particular epitope (similarly analogous to a key) on an antigen, allowing these two structures to bind together with precision. Using this binding mechanism, an antibody can *tag* a microbe or an infected cell for attack by other parts of the immune system, or can neutralize its target directly (for example, by blocking a part of a microbe that is essential for its invasion and survival). Depending on the antigen, the binding may impede the biological process causing the disease or may activate macrophages to destroy the foreign substance. The ability of an antibody to communicate with the other components of the immune system is mediated via its Fc region (located at the base of the "Y"), which contains a conserved glycosylation site involved in these interactions. The production of antibodies is the main function of the humoral immune system.

Antibodies are secreted by B cells of the adaptive immune system, mostly by differentiated B cells called plasma cells. Antibodies can occur in two physical forms, a soluble form that is secreted from the cell to be free in the blood plasma, and a membrane-bound form that is attached to the surface of a B cell and is referred to as the B-cell receptor (BCR). The BCR is found only on the surface of B cells and facilitates the activation of these cells and their subsequent differentiation into either antibody factories called plasma cells or memory B cells that will survive in the body and remember that same antigen so the B cells can respond faster upon future exposure. In most cases, interaction of the B cell with a T helper cell is necessary to produce full activation of the B cell and, therefore, antibody generation following antigen binding. Soluble antibodies are released into the blood and tissue fluids, as well as many secretions to continue to survey for invading microorganisms.

Antibodies are glycoproteins belonging to the immunoglobulin superfamily. They constitute most of the gamma globulin fraction of the blood proteins. They are typically made of basic structural units—each with two large heavy chains and two small light chains. There are several different types of antibody heavy chains that define the five different types of crystallisable fragments (Fc) that may be attached to the antigen-binding fragments. The five different types of Fc regions allow antibodies to be grouped into five *isotypes*. Each Fc region of a particular antibody isotype is able to bind to its specific Fc Receptor (except for IgD, which is essentially the BCR), thus allowing the antigen-antibody complex to mediate different roles depending on which FcR it binds. The ability of an antibody to bind to its corresponding FcR is further modulated by the structure of the glycan(s) present at conserved sites within its Fc region. The ability of antibodies to bind to FcRs helps to direct the appropriate immune response for each different type of foreign object they encounter. For example, IgE is responsible for an allergic response consisting of mast cell degranulation and histamine release. IgE's Fab paratope binds to allergic antigen, for example house dust mite particles, while its Fc region binds to Fc receptor ε. The allergen-IgE-FcRε interaction mediates allergic signal transduction to induce conditions such as asthma.

Though the general structure of all antibodies is very similar, a small region at the tip of the protein is extremely variable, allowing millions of antibodies with slightly different tip structures, or antigen-binding sites, to exist. This region is known as the *hypervariable region*. Each of these variants can bind to a different antigen. This enormous diversity of antibody paratopes on the antigen-binding fragments allows the immune system to recognize an equally wide variety of antigens. The large and diverse population of antibody paratope is generated by random recombination events of a set of gene segments that encode different antigen-binding sites (or *paratopes*), followed by random mutations in this area of the antibody gene, which create further diversity. This recombinational process that produces clonal antibody paratope diversity is called V(D)

J or VJ recombination. Basically, the antibody paratope is polygenic, made up of three genes, V, D, and J. Each paratope locus is also polymorphic, such that during antibody production, one allele of V, one of D, and one of J is chosen. These gene segments are then joined together using random genetic recombination to produce the paratope. The regions where the genes are randomly recombined together is the hyper variable region used to recognise different antigens on a clonal basis.

Antibody genes also re-organize in a process called class switching that changes the one type of heavy chain Fc fragment to another, creating a different isotype of the antibody that retains the antigen-specific variable region. This allows a single antibody to be used by different types of Fc receptors, expressed on different parts of the immune system.

Forms

The membrane-bound form of an antibody may be called a *surface immunoglobulin* (sIg) or a *membrane immunoglobulin* (mIg). It is part of the *B cell receptor* (BCR), which allows a B cell to detect when a specific antigen is present in the body and triggers B cell activation. The BCR is composed of surface-bound IgD or IgM antibodies and associated Ig-α and Ig-β heterodimers, which are capable of signal transduction. A typical human B cell will have 50,000 to 100,000 antibodies bound to its surface. Upon antigen binding, they cluster in large patches, which can exceed 1 micrometer in diameter, on lipid rafts that isolate the BCRs from most other cell signaling receptors. These patches may improve the efficiency of the cellular immune response. In humans, the cell surface is bare around the B cell receptors for several hundred nanometers, which further isolates the BCRs from competing influences.

Antibody–antigen Interactions

The antibody's paratope interacts with the antigen's epitope. An antigen usually contains different epitopes along its surface arranged discontinuously, and dominant epitopes on a given antigen are called determinants.

Antibody and antigen interact by spatial complementarity (lock and key). The molecular forces involved in the Fab-epitope interaction are weak and non-specific – for example electrostatic forces, hydrogen bonds, hydrophobic interactions, and van der Waals forces. This means binding between antibody and antigen is reversible, and the antibody's affinity towards an antigen is relative rather than absolute. Relatively weak binding also means it is possible for an antibody to cross-react with different antigens of different relative affinities.

Often, once an antibody and antigen bind, they become an immune complex, which functions as a unitary object and can act as an antigen in its own right, being countered by other antibodies. Similarly, haptens are small molecules that provoke no immune response by themselves, but once they bind to proteins, the resulting complex or hapten-carrier adduct is antigenic.

Isotypes

Antibodies can come in different varieties known as isotypes or classes. In placental mammals there are five antibody isotypes known as IgA, IgD, IgE, IgG, and IgM. They are each named with an "Ig" prefix that stands for immunoglobulin, a name sometimes used interchangeably with antibody, and differ in their biological properties, functional locations and ability to deal with different antigens, as depicted in the table. The different suffixes of the antibody isotypes denote the different types of heavy chains the antibody contains, with each heavy chain class named alphabetically: α (alpha), γ (gamma), δ (delta), ε (epsilon), and μ (mu). This gives rise to IgA, IgG, IgD, IgE, and IgM, respectively.

Antibody isotypes of mammals

Name	**Types**	**Description**	**Antibody complexes**
IgA	2	Found in mucosal areas, such as the gut, respiratory tract and urogenital tract, and prevents colonization by pathogens. Also found in saliva, tears, and breast milk.	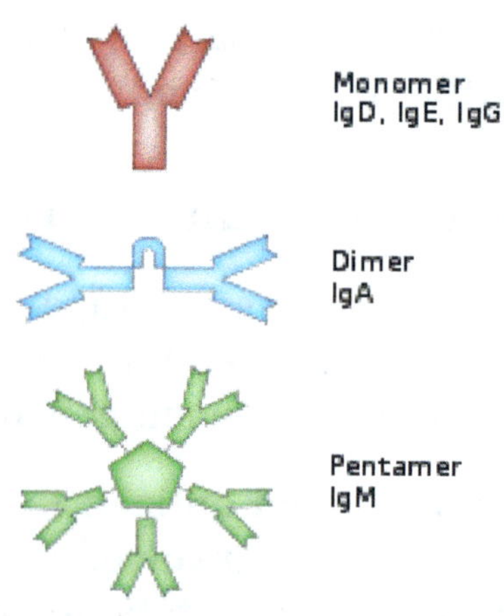
IgD	1	Functions mainly as an antigen receptor on B cells that have not been exposed to antigens. It has been shown to activate basophils and mast cells to produce antimicrobial factors.	
IgE	1	Binds to allergens and triggers histamine release from mast cells and basophils, and is involved in allergy. Also protects against parasitic worms.	
IgG	4	In its four forms, provides the majority of antibody-based immunity against invading pathogens. The only antibody capable of crossing the placenta to give passive immunity to the fetus.	
IgM	1	Expressed on the surface of B cells (monomer) and in a secreted form (pentamer) with very high avidity. Eliminates pathogens in the early stages of B cell-mediated (humoral) immunity before there is sufficient IgG.	

The antibody isotype of a B cell changes during cell development and activation. Immature B cells, which have never been exposed to an antigen, express only the IgM isotype in a cell surface bound form. The B lymphocyte, in this ready-to-respond form, is known as a "naive B lymphocyte." The naive B lymphocyte expresses both surface IgM and IgD. The co-expression of both of these immunoglobulin isotypes renders the B cell ready to respond to antigen. B cell activation follows engagement of the cell-bound antibody molecule with an antigen, causing the cell to divide and differentiate into an antibody-producing cell called a plasma cell. In this activated form, the B cell starts to produce antibody in a secreted form rather than a membrane-bound form. Some daughter cells of the activated B cells undergo isotype switching, a mechanism that causes the production of antibodies to change from IgM or IgD to the other antibody isotypes, IgE, IgA, or IgG, that have defined roles in the immune system.

Antibody isotypes not found in mammals		
Name	**Types**	**Description**
IgY		Found in birds and reptiles; related to mammalian IgG.
IgW		Found in sharks and skates; related to mammalian IgD.

Structure

Antibodies are heavy (~150 kDa) globularplasma proteins. They have sugar chains (glycans) added to conserved amino acid residues. In other words, antibodies are *glycoproteins*. The attached glycans are critically important to the structure and function of the antibody. Among other things the expressed glycans can modulate an antibody's affinity for its corresponding FcR(s).

The basic functional unit of each antibody is an immunoglobulin (Ig) monomer (containing only one Ig unit); secreted antibodies can also be dimeric with two Ig units as with IgA, tetrameric with four Ig units like teleost fish IgM, or pentameric with five Ig units, like mammalian IgM.

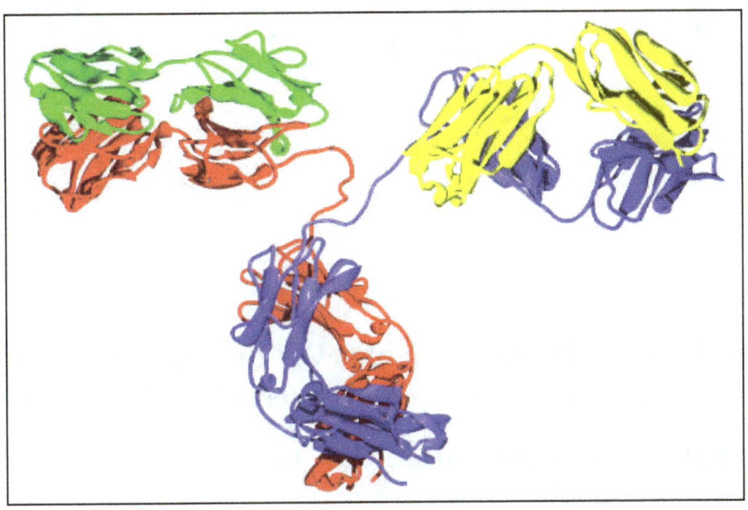

Several immunoglobulin domains make up the two heavy chains (red and blue) and the two light chains (green and yellow) of an antibody. The immunoglobulin domains are composed of between 7 (for constant domains) and 9 (for variable domains) β-strands.

The variable parts of an antibody are its V regions, and the constant part is its C region.

Immunoglobulin Domains

The Ig monomer is a "Y"-shaped molecule that consists of four polypeptide chains; two identical *heavy chains* and two identical *light chains* connected by disulfide bonds. Each chain is composed of structural domains called immunoglobulin domains. These domains contain about 70–110 amino acids and are classified into different categories (for example, variable or IgV, and constant or IgC) according to their size and function. They have a characteristic immunoglobulin fold in which two beta sheets create a "sandwich" shape, held together by interactions between conserved cysteines and other charged amino acids.

Heavy Chain

There are five types of mammalian Ig heavy chain denoted by the Greek letters: α, δ, ε, γ, and μ. The type of heavy chain present defines the *class* of antibody; these chains are found in IgA, IgD, IgE, IgG, and IgM antibodies, respectively. Distinct heavy chains differ in size and composition; α and γ contain approximately 450 amino acids, whereas μ and ε have approximately 550 amino acids.

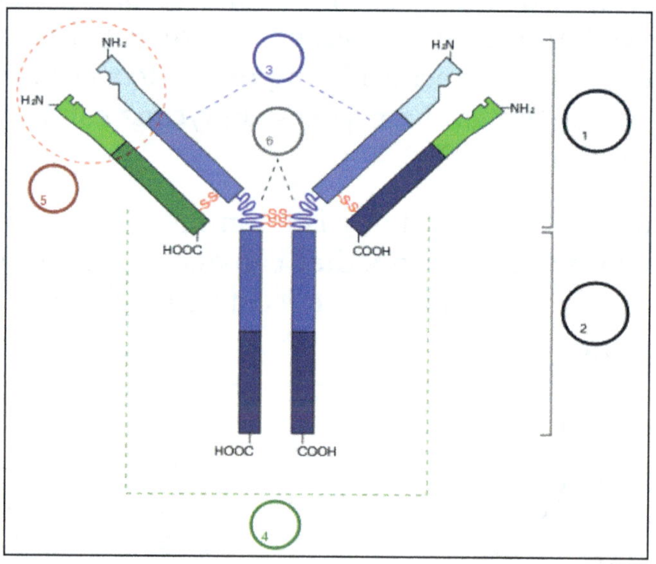

1. Fab region
2. Fc region
3. Heavy chain (blue) with one variable (V_H) domain followed by a constant domain (C_H1), a hinge region, and two more constant (C_H2 and C_H3) domains
4. Light chain (green) with one variable (V_L) and one constant (C_L) domain
5. Antigen binding site (paratope)
6. Hinge regions

Each heavy chain has two regions, the *constant region* and the *variable region.* The constant region is identical in all antibodies of the same isotype, but differs in antibodies of different isotypes. Heavy chains γ, α and δ have a constant region composed of *three* tandem (in a line) Ig domains, and a hinge region for added flexibility; heavy chains μ and ε have a constant region composed of *four* immunoglobulin domains. The variable region of the heavy chain differs in antibodies produced by different B cells, but is the same for all antibodies produced by a single B cell or B cell clone. The variable region of each heavy chain is approximately 110 amino acids long and is composed of a single Ig domain.

Light Chain

In mammals there are two types of immunoglobulin light chain, which are called lambda (λ) and kappa (κ). A light chain has two successive domains: one constant domain and one variable domain. The approximate length of a light chain is 211 to 217 amino acids. Each antibody contains two light chains that are always identical; only one type of light chain, κ or λ, is present per antibody in mammals. Other types of light chains, such as the iota (ι) chain, are found in other vertebrates like sharks (Chondrichthyes) and bony fishes (Teleostei).

CDRs, Fv, Fab and Fc Regions

Some parts of an antibody have the same functions. The arms of the Y, for example, contain the sites that can bind to antigens (in general, identical) and, therefore, recognize specific foreign objects. This region of the antibody is called the *Fab (fragment, antigen-binding) region*. It is composed of one constant and one variable domain from each heavy and light chain of the antibody. The paratope is shaped at the amino terminal end of the antibody monomer by the variable domains from the heavy and light chains. The variable domain is also referred to as the F_V region and is the most important region for binding to antigens. To be specific, variable loops of β-strands, three each on the light (V_L) and heavy (V_H) chains are responsible for binding to the antigen. These loops are referred to as the complementarity determining regions (CDRs). The structures of these CDRs have been clustered and classified by Chothia et al. and more recently by North et al. and Nikoloudis et al. In the framework of the immune network theory, CDRs are also called idiotypes. According to immune network theory, the adaptive immune system is regulated by interactions between idiotypes.

The base of the Y plays a role in modulating immune cell activity. This region is called the *Fc (Fragment, crystallizable) region*, and is composed of two heavy chains that contribute two or three constant domains depending on the class of the antibody. Thus, the Fc region ensures that each antibody generates an appropriate immune response for a given antigen, by binding to a specific class of Fc receptors, and other immune molecules, such as complement proteins. By doing this, it mediates different physiological effects including recognition of opsonized particles (binding to FcγR), lysis of cells (binding to complement), and degranulation of mast cells, basophils, and eosinophils (binding to FcεR).

In summary, the Fab region of the antibody determines antigen specificity while the Fc region of the antibody determines the antibody's class effect. Since only the constant domains of the heavy chains make up the Fc region of an antibody, the classes of heavy chain in antibodies determine their class effects. Possible classes of heavy chains in antibodies include alpha, gamma, delta, epsilon, and mu, and they define the antibody's isotypes IgA, G, D, E, and M, respectively. This infers different isotypes of antibodies have different class effects due to their different Fc regions binding and activating different types of receptors. Possible class effects of antibodies include: Opsonisation, agglutination, haemolysis, complement activation, mast cell degranulation, and neutralisation (though this class effect may be mediated by the Fab region rather than the Fc region). It also implies that Fab-mediated effects are directed at microbes or toxins, whilst Fc mediated effects are directed at effector cells or effector molecules.

Function

The main categories of antibody action include the following:

- Neutralisation, in which neutralizing antibodies block parts of the surface of a bacterial cell or virion to render its attack ineffective
- Agglutination, in which antibodies "glue together" foreign cells into clumps that are attractive targets for phagocytosis
- Precipitation, in which antibodies "glue together" serum-soluble antigens, forcing them to precipitate out of solution in clumps that are attractive targets for phagocytosis
- Complement activation (fixation), in which antibodies that are latched onto a foreign cell encourage complement to attack it with a membrane attack complex, which leads to the following:
    - Lysis of the foreign cell
    - Encouragement of inflammation by chemotactically attracting inflammatory cells

Activated B cells differentiate into either antibody-producing cells called plasma cells that secrete soluble antibody or memory cells that survive in the body for years afterward in order to allow the immune system to remember an antigen and respond faster upon future exposures.

At the prenatal and neonatal stages of life, the presence of antibodies is provided by passive immunization from the mother. Early endogenous antibody production varies for different kinds of antibodies, and usually appear within the first years of life. Since antibodies exist freely in the bloodstream, they are said to be part of the humoral immune system. Circulating antibodies are produced by clonal B cells that specifically respond to only one antigen (an example is a viruscapsid protein fragment). Antibodies contribute to immunity in three ways: They prevent pathogens from entering or damaging cells by binding to them; they stimulate removal of pathogens by macrophages and other cells by coating the pathogen; and they trigger destruction of pathogens by stimulating other immune responses such as the complement pathway. Antibodies will also trigger vasoactive amine degranulation to contribute to immunity against certain types of antigens (helminths, allergens).

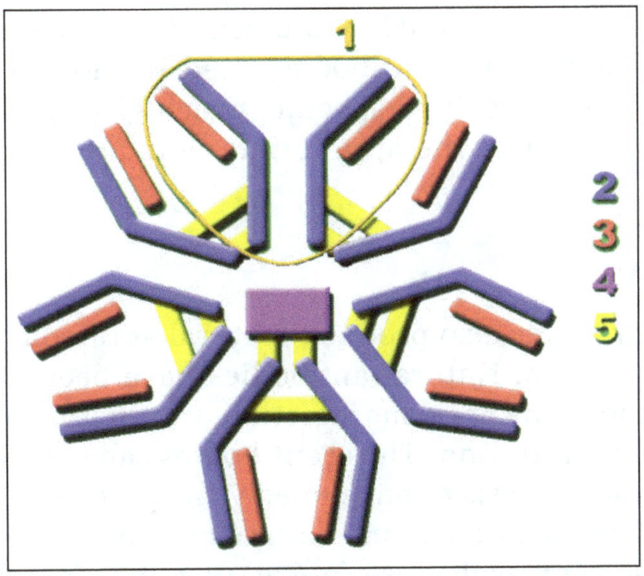

The secreted mammalian IgM has five Ig units. Each Ig unit (labeled 1) has two epitope binding Fab regions, so IgM is capable of binding up to 10 epitopes.

Activation of Complement

Antibodies that bind to surface antigens (for example, on bacteria) will attract the first component of the complement cascade with their Fc region and initiate activation of the "classical" complement system. This results in the killing of bacteria in two ways. First, the binding of the antibody and complement molecules marks the microbe for ingestion by phagocytes in a process called opsonization; these phagocytes are attracted by certain complement molecules generated in the complement cascade. Second, some complement system components form a membrane attack complex to assist antibodies to kill the bacterium directly (bacteriolysis).

Activation of Effector Cells

To combat pathogens that replicate outside cells, antibodies bind to pathogens to link them together, causing them to agglutinate. Since an antibody has at least two paratopes, it can bind more than one antigen by binding identical epitopes carried on the surfaces of these antigens. By coating the pathogen, antibodies stimulate effector functions against the pathogen in cells that recognize their Fc region.

Those cells that recognize coated pathogens have Fc receptors, which, as the name suggests, interact with the Fc region of IgA, IgG, and IgE antibodies. The engagement of a particular antibody with the Fc receptor on a particular cell triggers an effector function of that cell; phagocytes will phagocytose, mast cells and neutrophils will degranulate, natural killer cells will release cytokines and cytotoxic molecules; that will ultimately result in destruction of the invading microbe. The activation of natural killer cells by antibodies initiates a cytotoxic mechanism known as antibody-dependent cell-mediated cytotoxicity (ADCC) – this process may explain the efficacy of monoclonal antibodies used in biological therapies against cancer. The Fc receptors are isotype-specific, which gives greater flexibility to the immune system, invoking only the appropriate immune mechanisms for distinct pathogens.

Natural Antibodies

Humans and higher primates also produce "natural antibodies" that are present in serum before viral infection. Natural antibodies have been defined as antibodies that are produced without any previous infection, vaccination, other foreign antigen exposure or passive immunization. These antibodies can activate the classical complement pathway leading to lysis of enveloped virus particles long before the adaptive immune response is activated. Many natural antibodies are directed against the disaccharide galactose α(1,3)-galactose (α-Gal), which is found as a terminal sugar on glycosylated cell surface proteins, and generated in response to production of this sugar by bacteria contained in the human gut. Rejection of xenotransplantated organs is thought to be, in part, the result of natural antibodies circulating in the serum of the recipient binding to α-Gal antigens expressed on the donor tissue.

Immunoglobulin Diversity

Virtually all microbes can trigger an antibody response. Successful recognition and eradication of many different types of microbes requires diversity among antibodies; their amino acid composition varies allowing them to interact with many different antigens. It has been estimated that humans generate about 10 billion different antibodies, each capable of binding a distinct epitope of an antigen. Although a huge repertoire of different antibodies is generated in a single individual, the number of genes available to make these proteins is limited by the size of the human genome. Several complex genetic mechanisms have evolved that allow vertebrate B cells to generate a diverse pool of antibodies from a relatively small number of antibody genes.

Domain Variability

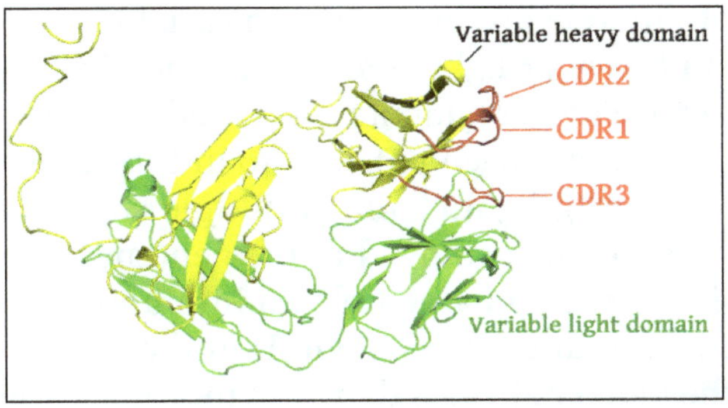

The complementarity determining regions of the heavy chain are shown in red (PDB: 1IGT).

The chromosomal region that encodes an antibody is large and contains several distinct gene loci for each domain of the antibody—the chromosome region containing heavy chain genes (IGH@) is found on chromosome 14, and the loci containing lambda and kappa light chain genes (IGL@ and IGK@) are found on chromosomes 22 and 2 in humans. One of these domains is called the variable domain, which is present in each heavy and light chain of every antibody, but can differ in different antibodies generated from distinct B cells. Differences, between the variable domains, are located on three loops known as hypervariable regions (HV-1, HV-2 and HV-3) or complementarity determining regions (CDR1, CDR2 and CDR3). CDRs are supported within the variable domains by conserved framework regions. The heavy chain locus contains about 65 different variable domain genes that all differ in their CDRs. Combining these genes with an array of genes for other domains of the antibody generates a large cavalry of antibodies with a high degree of variability. This combination is called V(D)J recombination discussed below.

V(D)J Recombination

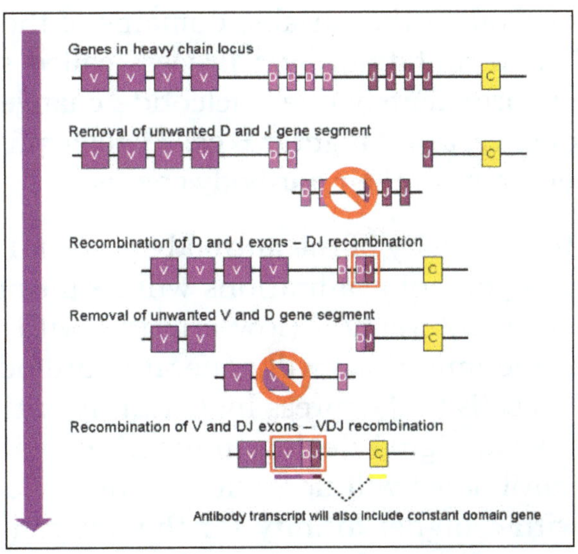

Simplified overview of V(D)J recombination of immunoglobulin heavy chains.

Somatic recombination of immunoglobulins, also known as *V(D)J recombination*, involves the generation of a unique immunoglobulin variable region. The variable region of each immunoglobulin heavy or light chain is encoded in several pieces—known as gene segments (subgenes). These segments are called variable (V), diversity (D) and joining (J) segments. V, D and J segments are found in Ig heavy chains, but only V and J segments are found in Ig light chains. Multiple copies of the V, D and J gene segments exist, and are tandemly arranged in the genomes of mammals. In the bone marrow, each developing B cell will assemble an immunoglobulin variable region by randomly selecting and combining one V, one D and one J gene segment (or one V and one J segment in the light chain). As there are multiple copies of each type of gene segment, and different combinations of gene segments can be used to generate each immunoglobulin variable region, this process generates a huge number of antibodies, each with different paratopes, and thus different antigen specificities. Interestingly, the rearrangement of several subgenes (i.e. V2 family) for lambda light chain immunoglobulin is coupled with the activation of microRNA miR-650, which further influences biology of B-cells.

RAG proteins play an important role with V(D)J recombination in cutting DNA at a particular region. Without the presence of these proteins, V(D)J recombination would not occur.

After a B cell produces a functional immunoglobulin gene during V(D)J recombination, it cannot express any other variable region (a process known as allelic exclusion) thus each B cell can produce antibodies containing only one kind of variable chain.

Somatic Hypermutation and Affinity Maturation

Following activation with antigen, B cells begin to proliferate rapidly. In these rapidly dividing cells, the genes encoding the variable domains of the heavy and light chains undergo a high rate of point mutation, by a process called *somatic hypermutation* (SHM). SHM results in approximately one nucleotide change per variable gene, per cell division. As a consequence, any daughter B cells will acquire slight amino acid differences in the variable domains of their antibody chains.

This serves to increase the diversity of the antibody pool and impacts the antibody's antigen-binding affinity. Some point mutations will result in the production of antibodies that have a weaker interaction (low affinity) with their antigen than the original antibody, and some mutations will generate antibodies with a stronger interaction (high affinity). B cells that express high affinity antibodies on their surface will receive a strong survival signal during interactions with other cells, whereas those with low affinity antibodies will not, and will die by apoptosis. Thus, B cells expressing antibodies with a higher affinity for the antigen will outcompete those with weaker affinities for function and survival. The process of generating antibodies with increased binding affinities is called *affinity maturation*. Affinity maturation occurs in mature B cells after V(D)J recombination, and is dependent on help from helper T cells.

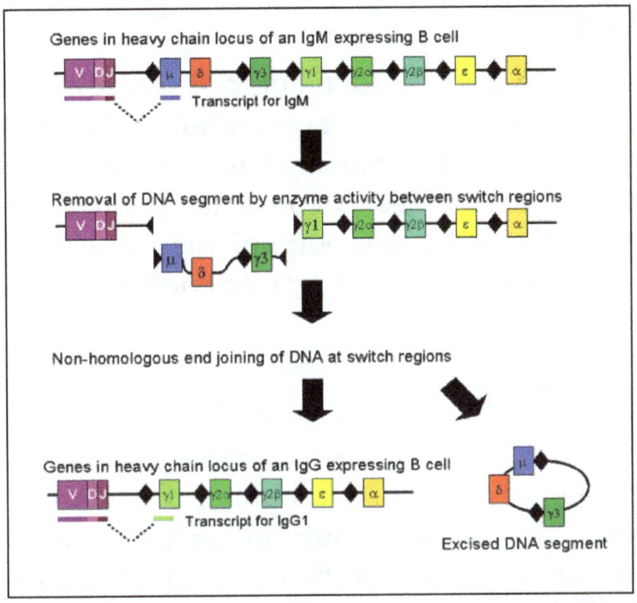

Mechanism of class switch recombination that allows isotype switching in activated B cells.

Class Switching

Isotype or class switching is a biological process occurring after activation of the B cell, which allows the cell to produce different classes of antibody (IgA, IgE, or IgG). The different classes of antibody, and thus effector functions, are defined by the constant (C) regions of the immunoglobulin heavy chain. Initially, naive B cells express only cell-surface IgM and IgD with identical antigen binding regions. Each isotype is adapted for a distinct function; therefore, after activation, an antibody with an IgG, IgA, or IgE effector function might be required to effectively eliminate an antigen. Class switching allows different daughter cells from the same activated B cell to produce antibodies of different isotypes. Only the constant region of the antibody heavy chain changes during class switching; the variable regions, and therefore antigen specificity, remain unchanged. Thus the progeny of a single B cell can produce antibodies, all specific for the same antigen, but with the ability to produce the effector function appropriate for each antigenic challenge. Class switching is triggered by cytokines; the isotype generated depends on which cytokines are present in the B cell environment.

Class switching occurs in the heavy chain gene locus by a mechanism called class switch recombination (CSR). This mechanism relies on conserved nucleotide motifs, called *switch (S) regions*, found in DNA upstream of each constant region gene (except in the δ-chain). The DNA strand is broken by the activity of a series of enzymes at two selected

S-regions. The variable domain exon is rejoined through a process called non-homologous end joining (NHEJ) to the desired constant region (γ, α or ε). This process results in an immunoglobulin gene that encodes an antibody of a different isotype.

Affinity Designations

A group of antibodies can be called *monovalent* (or *specific*) if they have affinity for the same epitope, or for the same antigen (but potentially different epitopes on the molecule), or for the same strain of microorganism (but potentially different antigens on or in it). In contrast, a group of antibodies can be called *polyvalent* (or *unspecific*) if they have affinity for various antigens or microorganisms.Intravenous immunoglobulin, if not otherwise noted, consists of polyvalent IgG. In contrast, monoclonal antibodies are monovalent for the same epitope.

Asymmetrical Antibodies

Heterodimeric antibodies, which are also asymmetrical and antibodies, allow for greater flexibility and new formats for attaching a variety of drugs to the antibody arms. One of the general formats for a heterodimeric antibody is the "knobs-into-holes" format. This format is specific to the heavy chain part of the constant region in antibodies. The "knobs" part is engineered by replacing a small amino acid with a larger one. It fits into the "hole", which is engineered by replacing a large amino acid with a smaller one. What connects the "knobs" to the "holes" are the disulfide bonds between each chain. The "knobs-into-holes" shape facilitates antibody dependent cell mediated cytotoxicity. Single chain variable fragments (scFv) are connected to the variable domain of the heavy and light chain via a short linker peptide. The linker is rich in glycine, which gives it more flexibility, and serine/threonine, which gives it specificity. Two different scFv fragments can be connected together, via a hinge region, to the constant domain of the heavy chain or the constant domain of the light chain. This gives the antibody bispecificity, allowing for the binding specificities of two different antigens. The "knobs-into-holes" format enhances heterodimer formation but doesn't suppress homodimer formation.

To further improve the function of heterodimeric antibodies, many scientists are looking towards artificial constructs. Artificial antibodies are largely diverse protein motifs that use the functional strategy of the antibody molecule, but aren't limited by the loop and framework structural constraints of the natural antibody. Being able to control the combinational design of the sequence and three-dimensional space could transcend the natural design and allow for the attachment of different combinations of drugs to the arms.

Heterodimeric antibodies have a greater range in shapes they can take and the drugs that are attached to the arms don't have to be the same on each arm, allowing for different combinations of drugs to be used in cancer treatment. Pharmaceuticals are able to produce highly functional bispecific, and even multispecific, antibodies. The degree to which they can function is impressive given that such a change shape from the natural form should lead to decreased functionality.

Medical Applications

Disease Diagnosis

Detection of particular antibodies is a very common form of medical diagnostics, and applications such as serology depend on these methods. For example, in biochemical assays for disease diagnosis, a titer of antibodies directed against Epstein-Barr virus or Lyme disease is estimated from the blood. If those antibodies are not present, either the person is not infected or the infection occurred a *very* long time ago, and the B cells generating these specific antibodies have naturally decayed.

In clinical immunology, levels of individual classes of immunoglobulins are measured by nephelometry (or turbidimetry) to characterize the antibody profile of patient. Elevations in different classes of immunoglobulins are sometimes useful in determining the cause of liver damage in patients for whom the diagnosis is unclear. For example, elevated IgA indicates alcoholic cirrhosis, elevated IgM indicates viral hepatitis and primary biliary cirrhosis, while IgG is elevated in viral hepatitis, autoimmune hepatitis and cirrhosis.

Autoimmune disorders can often be traced to antibodies that bind the body's own epitopes; many can be detected through blood tests. Antibodies directed against red blood cell surface antigens in immune mediated hemolytic anemia are detected with the Coombs test. The Coombs test is also used for antibody screening in blood transfusion preparation and also for antibody screening in antenatal women.

Practically, several immunodiagnostic methods based on detection of complex antigen-antibody are used to diagnose infectious diseases, for example ELISA, immunofluorescence, Western blot, immunodiffusion, immunoelectrophoresis, and magnetic immunoassay. Antibodies raised against human chorionic gonadotropin are used in over the counter pregnancy tests.

New dioxaborolane chemistry enables radioactive fluoride (^{18}F) labeling of antibodies, which allows for positron emission tomography (PET) imaging of cancer.

Disease Therapy

Targeted monoclonal antibody therapy is employed to treat diseases such as rheumatoid arthritis,multiple sclerosis,psoriasis, and many forms of cancer including non-Hodgkin's lymphoma,colorectal cancer, head and neck cancer and breast cancer.

Some immune deficiencies, such as X-linked agammaglobulinemia and hypogammaglobulinemia, result in partial or complete lack of antibodies. These diseases are often treated by inducing a short term form of immunity called passive immunity. Passive immunity is achieved through the transfer of ready-made antibodies in the form of human or animal serum, pooled immunoglobulin or monoclonal antibodies, into the affected individual.

Prenatal Therapy

Rhesus factor, also known as Rhesus D (RhD) antigen, is an antigen found on red blood cells; individuals that are Rhesus-positive (Rh+) have this antigen on their red blood cells and individuals that are Rhesus-negative (Rh–) do not. During normal childbirth, delivery trauma or complications during pregnancy, blood from a fetus can enter the mother's system. In the case of an Rh-incompatible mother and child, consequential blood mixing may sensitize an Rh- mother to the Rh antigen on the blood cells of the Rh+ child, putting the remainder of the pregnancy, and any subsequent pregnancies, at risk for hemolytic disease of the newborn.

Rho(D) immune globulin antibodies are specific for human Rhesus D (RhD) antigen. Anti-RhD antibodies are administered as part of a prenatal treatment regimen to prevent sensitization that may occur when a Rhesus-negative mother has a Rhesus-positive fetus. Treatment of a mother with Anti-RhD antibodies prior to and immediately after trauma and delivery destroys Rh antigen in the mother's system from the fetus. It is important to note that this occurs before the antigen can stimulate maternal B cells to "remember" Rh antigen by generating memory B cells. Therefore, her humoral immune system will not make anti-Rh antibodies, and will not attack the Rhesus antigens of the current or subsequent babies. Rho(D) Immune Globulin treatment prevents sensitization that can lead to Rh disease, but does not prevent or treat the underlying disease itself.

Research Applications

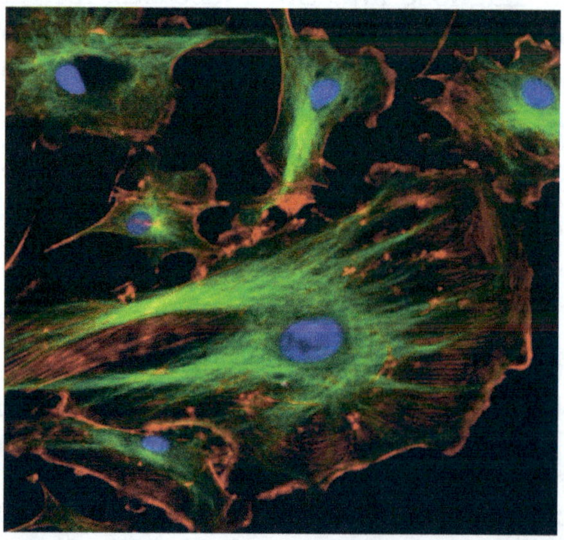

Immunofluorescence image of the eukaryotic cytoskeleton.
Actin filaments are shown in red, microtubules in green, and the nuclei in blue.

Specific antibodies are produced by injecting an antigen into a mammal, such as a mouse, rat, rabbit, goat, sheep, or horse for large quantities of antibody. Blood isolated from these animals contains *polyclonal antibodies*—multiple antibodies that bind to the same antigen—in the serum, which can now be called antiserum. Antigens are also injected into chickens for generation of polyclonal antibodies in egg yolk. To obtain antibody that is specific for a single epitope of an antigen, antibody-secreting lymphocytes are isolated from the animal and immortalized by fusing them with a cancer cell line. The fused cells are called hybridomas, and will continually grow and secrete antibody in culture. Single hybridoma cells are isolated by dilution cloning to generate cell clones that all produce the same antibody; these antibodies are called *monoclonal antibodies*. Polyclonal and monoclonal antibodies are often purified using Protein A/G or antigen-affinity chromatography.

In research, purified antibodies are used in many applications. Antibodies for research applications can be found directly from antibody suppliers, or through use of a specialist search engine. Research antibodies are most commonly used to identify and locate intracellular and extracellular proteins. Antibodies are used in flow cytometry to differentiate cell types by the proteins they express; different types of cell express different combinations of cluster of differentiation molecules on their surface, and produce different intracellular and secretable proteins. They are also used in immunoprecipitation to separate proteins and anything bound to them (co-immunoprecipitation) from other molecules in a cell lysate, in Western blot analyses to identify proteins separated by electrophoresis, and in immunohistochemistry or immunofluorescence to examine protein expression in tissue sections or to locate proteins within cells with the assistance of a microscope. Proteins can also be detected and quantified with antibodies, using ELISA and ELISPOT techniques.

- Reproducibility in Science

Antibodies used in research are some of the most powerful, yet most problematic reagents with a tremendous number of factors that must be controlled in any experiment including cross reactivity, or the antibody recognizing multiple epitopes and affinity, which can vary widely different depending on experimental conditions such as pH, solvent, state of tissue etc. Multiple attempts have been made to improve both the way that researchers validate antibodies and ways in which they report on antibodies. Researchers using antibodies in their work need to record them correctly in order to allow their research to be reproducible (and therefore tested, and qualified by other researchers). Less than half of research antibodies referenced in academic papers can be easily identified. Papers published in F1000 in 2014 and 2015 provide researchers with a guide for reporting research antibody use. The RRID paper, is co-published in 4 journals that implemented the RRIDs Standard for research resource citation, which draws data from the antibodyregistry.org as the source of antibody identifiers.

Regulatory Validation of Monoclonal Antibody Products for Human Use

Production and testing :

Traditionally, most antibodies are produced by hybridoma cell lines through immortalization of antibody-producing cells by chemically-induced fusion with myeloma cells. In some cases, additional fusions with other lines have created "triomas" and "quadromas". The manufacturing process should be appropriately described and validated. Validation studies should at least include :

- The demonstration that the process is able to produce in good quality (the process should be validated)
- The efficiency of the antibody purification (all impurities and virus must be eliminated)
- The characterization of purified antibody (physicochemical characterization, immunological properties, biological activities, contaminants, ...)
- Determination of the virus clearance studies

Before clinical trials, studies of product safety and feasibility have to be performed :

- Product safety testing : Sterility (bacteria and fungi), In vitro and in vivo testing for adventitious viruses, Murine retrovirus testing... Product safety data needed before the initiation of feasibility trials in serious or immediately life-threatening conditions, it serves to evaluate dangerous potential of the product.
- Feasibility testing : These are pilot studies whose objectives include, among others, early characterization of safety and initial proof of concept in a small specific patient population (in vito or in vivo testing).

Preclinical studies :

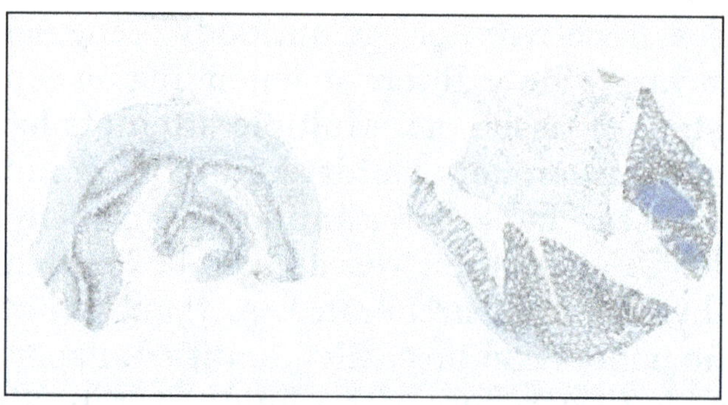

Example of tissue cross-reactivity on human tissue microarray.

- Testing cross-reactivity of antibody : to highlight unwanted interactions (toxicity) of antibodies with previously characterized tissues. This study can be performed in vitro (Reactivity of the antibody or immunoconjugate should be determined with a quick-frozen adult tissues) or in vivo (with appropriates animal models). More informations about in vitro cross-reactivity testing.

- Preclinical pharmacology and toxicity testing : Preclinical safety testing of antibody is designed to identify possible toxicities in humans, to estimate the likelihood and severity of potential adverse events in humans, and to identify a safe starting dose and dose escalation, when possible.
- Animal toxicity studies : Acute toxicity testing, Repeat-dose toxicity testing, Long-term toxicity testing http://www.animalresearch.info/en/drug-development/safety-testing/
- Pharmacokinetics and pharmacodynamics testing : Use for determinate clinical dosages, antibody activities (AUC, pharmacodynamics, biodistribution, ...), evaluation of the potential clinical effects.

Structure Prediction

The importance of antibodies in health care and the biotechnology industry demands knowledge of their structures at high resolution. This information is used for protein engineering, modifying the antigen binding affinity, and identifying an epitope, of a given antibody. X-ray crystallography is one commonly used method for determining antibody structures. However, crystallizing an antibody is often laborious and time-consuming. Computational approaches provide a cheaper and faster alternative to crystallography, but their results are more equivocal, since they do not produce empirical structures. Online web servers such as *Web Antibody Modeling* (WAM) and *Prediction of Immunoglobulin Structure* (PIGS) enables computational modeling of antibody variable regions. Rosetta Antibody is a novel antibody F_V region structure prediction server, which incorporates sophisticated techniques to minimize CDR loops and optimize the relative orientation of the light and heavy chains, as well as homology models that predict successful docking of antibodies with their unique antigen.

The ability to describe the antibody through binding affinity to the antigen is supplemented by information on antibody structure and amino acid sequences for the purpose of patent claims.

History

The first use of the term "antibody" occurred in a text by Paul Ehrlich. The term *Antikörper* (the German word for *antibody*) appears in the conclusion of his article "Experimental Studies on Immunity", published in October 1891, which states that, "if two substances give rise to two different antikörper, then they themselves must be different". However, the term was not accepted immediately and several other terms for antibody were proposed; these included *Immunkörper*, *Amboceptor*, *Zwischenkörper*, *substance sensibilisatrice*, *copula*, *Desmon*, *philocytase*, *fixateur*, and *Immunisin*. The word *antibody* has formal analogy to the word *antitoxin* and a similar concept to *Immunkörper* (*immune body* in English). As such, the original construction of the word contains a logical flaw; the antitoxin is something directed against a toxin, while the antibody is a body directed against something.

Angel of the West (2008) by Julian Voss-Andreae is a sculpture based on the antibody structure published by E. Padlan. Created for the Florida campus of the Scripps Research Institute, the antibody is placed into a ring referencing Leonardo da Vinci's *Vitruvian Man* thus highlighting the similarity of the antibody and the human body.

The study of antibodies began in 1890 when Kitasato Shibasaburō described antibody activity against diphtheria and tetanus toxins. Kitasato put forward the theory of humoral immunity, proposing that a mediator in serum could react with a foreign antigen. His idea prompted Paul Ehrlich to propose the side-chain theory for antibody and antigen interaction in 1897, when he hypothesized that receptors (described as "side-chains") on the surface of cells could bind specifically to toxins – in a «lock-and-key» interaction – and that this binding reaction is the trigger for the production of antibodies. Other researchers believed that antibodies existed freely in the blood and, in 1904, Almroth Wright suggested that soluble antibodies coated bacteria to label them for phagocytosis and killing; a process that he named opsoninization.

Michael Heidelberger.

In the 1920s, Michael Heidelberger and Oswald Avery observed that antigens could be precipitated by antibodies and went on to show that antibodies are made of protein. The biochemical properties of antigen-antibody-binding interactions were examined in more detail in the late 1930s by John Marrack. The next major advance was in the 1940s, when Linus Pauling confirmed the lock-and-key theory proposed by Ehrlich by showing that the interactions between antibodies and antigens depend more on their shape than their chemical composition. In 1948, Astrid Fagreaus discovered that B cells, in the form of plasma cells, were responsible for generating antibodies.

Further work concentrated on characterizing the structures of the antibody proteins. A major advance in these structural studies was the discovery in the early 1960s by Gerald Edelman and Joseph Gally of the antibody light chain, and their realization that this protein is the same as the Bence-Jones protein described in 1845 by Henry Bence Jones. Edelman went on to discover that antibodies are composed of disulfide bond-linked heavy and light chains. Around the same time, antibody-binding (Fab) and antibody tail (Fc) regions of IgG were characterized by Rodney Porter. Together, these scientists deduced the structure and complete amino acid sequence of IgG, a feat for which they were jointly awarded the 1972 Nobel Prize in Physiology or Medicine. The Fv fragment was prepared and characterized by David Givol. While most of these early studies focused on IgM and IgG, other immunoglobulin isotypes were identified in the 1960s: Thomas Tomasi discovered secretory antibody (IgA); David S. Rowe and John L. Fahey discovered IgD; and Kimishige Ishizaka and Teruko Ishizaka discovered IgE and showed it was a class of antibodies involved in allergic reactions. In a landmark series of experiments beginning in 1976, Susumu Tonegawa showed that genetic material can rearrange itself to form the vast array of available antibodies.

Antibody Mimetic

Antibody mimetics are organic compounds that, like antibodies, can specifically bind antigens. They are usually artificial peptides or proteins with a molar mass of about 3 to 20 kDa. Nucleic acids and small molecules are sometimes considered antibody mimetics, but not artificial antibodies, antibody fragments and fusion proteins are composed from these. Common advantages over antibodies are better solubility, tissue penetration, stability towards heat and enzymes, and comparatively low production costs. Antibody mimetics such as the Affimer and the DARPin have being developed and commercialised as research, diagnostic and therapeutic agents.

Collagen

Collagen is the main structural protein in the extracellular space in the various connective tissues in animal bodies. As the main component of connective tissue, it is the most abundant protein in mammals, making up from 25% to 35% of the whole-body protein content. Depending upon the degree of mineralization, collagen tissues may be rigid (bone), compliant (tendon), or have a gradient from rigid to compliant (cartilage). Collagen, in the form of elongated fibrils, is mostly found in fibrous tissues such as tendons, ligaments and skin. It is also abundant in corneas, cartilage, bones, blood vessels, the gut, intervertebral discs and the dentin in teeth. In muscle tissue, it serves as a major component of the endomysium. Collagen constitutes one to two percent of muscle tissue, and accounts for 6% of the weight of strong, tendinous muscles. The fibroblast is the most common cell that creates collagen.

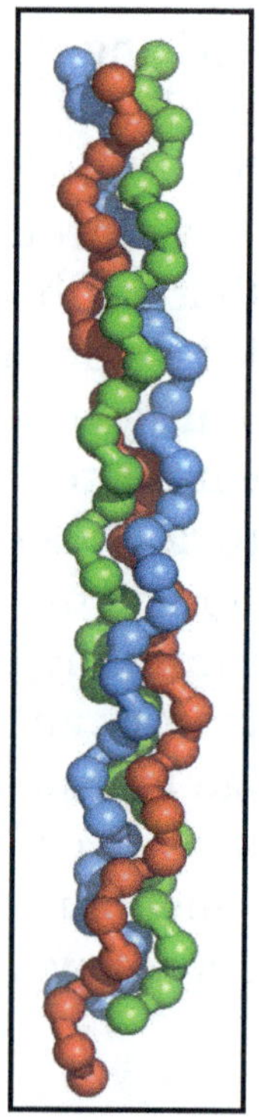

Tropocollagen molecule: three left-handed procollagens (red, green, blue) join to form a right handed triple helical tropocollagen.

Gelatin, which is used in food and industry, is collagen that has been irreversibly hydrolyzed. Collagen also has many medical uses in treating complications of the bones and skin.

Types of Collagen

Collagen occurs in many places throughout the body. Over 90% of the collagen in the human body, however, is type I.

So far, 28 types of collagen have been identified and described. They can be divided into several groups according to the structure they form:

- Fibrillar (Type I, II, III, V, XI)
- Non-fibrillar

- FACIT (Fibril Associated Collagens with Interrupted Triple Helices) (Type IX, XII, XIV, XVI, XIX)
- Short chain (Type VIII, X)
- Basement membrane (Type IV)
- Multiplexin (Multiple Triple Helix domains with Interruptions) (Type XV, XVIII)
- MACIT (Membrane Associated Collagens with Interrupted Triple Helices) (Type XIII, XVII)
- Other (Type VI, VII)

The five most common types are:

- Type I: skin, tendon, vascular ligature, organs, bone (main component of the organic part of bone).
- Type II: cartilage (main collagenous component of cartilage).
- Type III: reticulate (main component of reticular fibers), commonly found alongside type I.
- Type IV: forms basal lamina, the epithelium-secreted layer of the basement membrane.
- Type V: cell surfaces, hair and placenta.

Medical uses

Cardiac Applications

The collagenous cardiac skeleton which includes the four heart valve rings, is histologically, elastically and uniquely bound to cardiac muscle. The cardiac skeleton also includes the separating septa of the heart chambers – the interventricular septum and the atrioventricular septum. Collagen contribution to the measure of cardiac performance summarily represents a continuous torsional force opposed to the fluid mechanics of blood pressure emitted from the heart. The collagenous structure that divides the upper chambers of the heart from the lower chambers is an impermeable membrane that excludes both blood and electrical impulses through typical physiological means. With support from collagen, atrial fibrillation should never deteriorate to ventricular fibrillation. Collagen is layered in variable densities with cardiac muscle mass. The mass, distribution, age and density of collagen all contribute to the compliance required to move blood back and forth. Individual cardiac valvular leaflets are folded into shape by specialized collagen under variable pressure. Gradual calcium deposition within collagen occurs as a natural function of aging. Calcified points within collagen matrices show contrast in a moving display of blood and muscle, enabling methods of cardiac imaging technology to arrive at ratios essentially stating blood in (cardiac input) and blood out (cardiac output). Pathology of the collagen underpinning of the heart is understood within the category of connective tissue disease.

Cosmetic Surgery

Collagen has been widely used in cosmetic surgery, as a healing aid for burn patients for reconstruction of bone and a wide variety of dental, orthopedic, and surgical purposes. Both human and bovine collagen is widely used as dermal fillers for treatment of wrinkles and skin aging. Some points of interest are:

1. When used cosmetically, there is a chance of allergic reactions causing prolonged redness; however, this can be virtually eliminated by simple and inconspicuous patch testing prior to cosmetic use.

2. Most medical collagen is derived from young beef cattle (bovine) from certified BSE-free animals. Most manufacturers use donor animals from either "closed herds", or from countries which have never had a reported case of BSE such as Australia, Brazil, and New Zealand.

Bone Grafts

As the skeleton forms the structure of the body, it is vital that it maintains its strength, even after breaks and injuries. Collagen is used in bone grafting as it has a triple helical structure, making it a very strong molecule. It is ideal for use in bones, as it does not compromise the structural integrity of the skeleton. The triple helical structure of collagen prevents it from being broken down by enzymes, it enables adhesiveness of cells and it is important for the proper assembly of the extracellular matrix.

Tissue Regeneration

Collagen scaffolds are used in tissue regeneration, whether in sponges, thin sheets, or gels. Collagen has the correct properties for tissue regeneration such as pore structure, permeability, hydrophilicity and it is stable in vivo. Collagen scaffolds are also ideal for the deposition of cells, such as osteoblasts and fibroblasts and once inserted, growth is able to continue as normal in the tissue.

Reconstructive Surgical uses

Collagens are widely employed in the construction of the artificial skin substitutes used in the management of severe burns. These collagens may be derived from bovine, equine, porcine, or even human sources; and are sometimes used in combination with silicones, glycosaminoglycans, fibroblasts, growth factors and other substances.

Collagen is also sold commercially in pill form as a supplement to aid joint mobility. However, because proteins are broken down into amino acids before absorption, there is no reason for orally ingested collagen to affect connective tissue in the body, except through the effect of individual amino acid supplementation.

Collagen is also frequently used in scientific research applications for cell culture, studying cell behavior and cellular interactions with the extracellular environment.

Wound Care

Collagen is one of the body's key natural resources and a component of skin tissue that can benefit all stages of the wound healing process. When collagen is made available to the wound bed, closure can occur. Wound deterioration, followed sometimes by procedures such as amputation, can thus be avoided.

Collagen is a natural product, therefore it is used as a natural wound dressing and has properties that artificial wound dressings do not have. It is resistant against bacteria, which is of vital importance in a wound dressing. It helps to keep the wound sterile, because of its natural ability to fight infection. When collagen is used as a burn dressing, healthy granulation tissue is able to form very quickly over the burn, helping it to heal rapidly.

Throughout the 4 phases of wound healing, collagen performs the following functions in wound healing:

- Guiding function: Collagen fibers serve to guide fibroblasts. Fibroblasts migrate along a connective tissue matrix.
- Chemotactic properties: The large surface area available on collagen fibers can attract fibrogenic cells which help in healing.
- Nucleation: Collagen, in the presence of certain neutral salt molecules can act as a nucleating agent causing formation of fibrillar structures. A collagen wound dressing might serve as a guide for orienting new collagen deposition and capillary growth.
- Hemostatic properties: Blood platelets interact with the collagen to make a hemostatic plug.

Veterinary Use

Some studies have shown efficacy of collagen supplementation for dogs with osteoarthritis pain, alone or in combination with other nutraceuticals like glucosamine and chondroitin.

Chemistry

The collagen protein is composed of a triple helix, which generally consists of two identical chains (α1) and an additional chain that differs slightly in its chemical composition (α2). The amino acid composition of collagen is atypical for proteins, particularly with respect to its high hydroxyproline content. The most common motifs in the amino acid sequence of collagen are glycine-proline-X and glycine-X-hydroxyproline, where X is any amino acid other than glycine, proline or hydroxyproline. The average amino acid composition for fish and mammal skin is given.

Amino acid	Abundance in mammal skin (residues/1000)	Abundance in fish skin (residues/1000)
Glycine	329	339
Proline	126	108
Alanine	109	114
Hydroxyproline	95	67
Glutamic acid	74	76
Arginine	49	52
Aspartic acid	47	47
Serine	36	46
Lysine	29	26
Leucine	24	23
Valine	22	21
Threonine	19	26
Phenylalanine	13	14
Isoleucine	11	11
Hydroxylysine	6	8
Methionine	6	13
Histidine	5	7
Tyrosine	3	3
Cysteine	1	1
Tryptophan	0	0

Synthesis

First, a three-dimensional stranded structure is assembled, with the amino acids glycine and proline as its principal components. This is not yet collagen but its precursor, procollagen. Procollagen is then modified by the addition of hydroxyl groups to the amino acids proline and lysine. This step is important for later glycosylation and the formation of the triple helix structure of collagen. The hydroxylase enzymes that perform these reactions require Vitamin C as a cofactor, and a deficiency in this vitamin results in impaired collagen synthesis and the resulting disease scurvy These hydroxylation reactions are catalyzed by two different enzymes: prolyl-4-hydroxylase and lysyl-hydroxylase. Vitamin C also serves with them in inducing these reactions. In this service, one molecule of vitamin C is destroyed for each H replaced by OH. The synthesis of collagen occurs inside and outside of the cell. The formation of collagen which results in fibrillary collagen (most common form) is discussed here. Meshwork collagen, which is often involved in the formation of filtration systems, is the other form of collagen. All types of collagens are triple helices, and the differences lie in the make-up of the alpha peptides created in step 2.

1. Transcription of mRNA: About 34 genes are associated with collagen formation, each coding for a specific mRNA sequence, and typically have the "*COL*" prefix. The beginning of collagen synthesis begins with turning on genes which are associated with the formation of a particular alpha peptide (typically alpha 1, 2 or 3).

2. Pre-pro-peptide formation: Once the final mRNA exits from the cell nucleus and enters into the cytoplasm, it links with the ribosomal subunits and the process of translation occurs. The early/first part of the new peptide is known as the signal sequence. The signal sequence on the N-terminal of the peptide is recognized by a signal recognition particle on the endoplasmic reticulum, which will be responsible for directing the pre-pro-peptide into the endoplasmic reticulum. Therefore, once the synthesis of new peptide is finished, it goes directly into the endoplasmic reticulum for post-translational processing. It is now known as pre-pro-collagen.

3. Pre-pro-peptide to pro-collagen: Three modifications of the pre-pro-peptide occur leading to the formation of the alpha peptide:

    1. The signal peptide on the N-terminal is dissolved, and the molecule is now known as *propeptide* (not procollagen).

    2. Hydroxylation of lysines and prolines on propeptide by the enzymes 'prolyl hydroxylase' and 'lysyl hydroxylase' (to produce hydroxyproline and hydroxylysine) occurs to aid cross-linking of the alpha peptides. This enzymatic step requires vitamin C as a cofactor. In scurvy, the lack of hydroxylation of prolines and lysines causes a looser triple helix (which is formed by three alpha peptides).

    3. Glycosylation occurs by adding either glucose or galactose monomers onto the hydroxyl groups that were placed onto lysines, but not on prolines.

    4. Once these modifications have taken place, three of the hydroxylated and glycosylated propeptides twist into a triple helix forming procollagen. Procollagen still has unwound ends, which will be later trimmed. At this point, the procollagen is packaged into a transfer vesicle destined for the Golgi apparatus.

4. Golgi apparatus modification: In the Golgi apparatus, the procollagen goes through one last post-translational modification before being secreted out of the cell. In this step, oligosaccharides (not monosaccharides as in step 3) are added, and then the procollagen is packaged into a secretory vesicle destined for the extracellular space.

5. Formation of tropocollagen: Once outside the cell, membrane bound enzymes known as 'collagen peptidases', remove the "loose ends" of the procollagen molecule. What is left is known as tropocollagen. Defects in this step produce one of the many collagenopathies known as Ehlers-Danlos syndrome. This step is absent when synthesizing type III, a type of fibrilar collagen.

6. Formation of the collagen fibril: 'Lysyl oxidase', an extracellular enzyme, produces the final step in the collagen synthesis pathway. This enzyme acts on lysines and hydroxylysines producing aldehyde groups, which will eventually undergo covalent bonding between tropocollagen molecules. This polymer of tropocollogen is known as a collagen fibril.

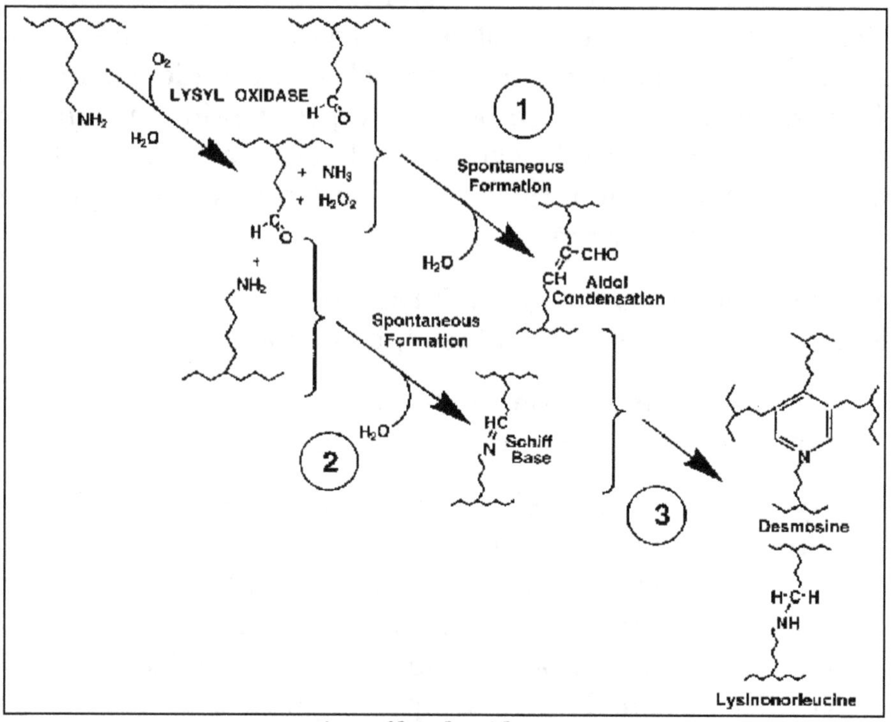

Action of lysyl oxidase.

Amino Acids

Collagen has an unusual amino acid composition and sequence:

- Glycine is found at almost every third residue.
- Proline makes up about 17% of collagen.

- Collagen contains two uncommon derivative amino acids not directly inserted during translation. These amino acids are found at specific locations relative to glycine and are modified post-translationally by different enzymes, both of which require vitamin C as a cofactor.
    - Hydroxyproline derived from proline.
    - Hydroxylysine derived from lysine - depending on the type of collagen, varying numbers of hydroxylysines are glycosylated (mostly having disaccharides attached).

Cortisol stimulates degradation of (skin) collagen into amino acids.

Collagen I Formation

Most collagen forms in a similar manner, but the following process is typical for type I:

1. Inside the cell;
    1. Two types of alpha chains are formed during translation on ribosomes along the rough endoplasmic reticulum (RER): alpha-1 and alpha-2 chains. These peptide chains (known as preprocollagen) have registration peptides on each end and a signal peptide.
    2. Polypeptide chains are released into the lumen of the RER.
    3. Signal peptides are cleaved inside the RER and the chains are now known as pro-alpha chains.
    4. Hydroxylation of lysine and proline amino acids occurs inside the lumen. This process is dependent on ascorbic acid (vitamin C) as a cofactor.
    5. Glycosylation of specific hydroxylysine residues occurs.
    6. Triple alpha helical structure is formed inside the endoplasmic reticulum from two alpha-1 chains and one alpha-2 chain.
    7. Procollagen is shipped to the Golgi apparatus, where it is packaged and secreted by exocytosis.
2. Outside the cell
    1. Registration peptides are cleaved and tropocollagen is formed by procollagen peptidase.
    2. Multiple tropocollagen molecules form collagen fibrils, via covalent cross-linking (aldol reaction) by lysyl oxidase which links hydroxylysine and lysine residues. Multiple collagen fibrils form into collagen fibers.

3. Collagen may be attached to cell membranes via several types of protein, including fibronectin and integrin.

Synthetic Pathogenesis

Vitamin C deficiency causes scurvy, a serious and painful disease in which defective collagen prevents the formation of strong connective tissue. Gums deteriorate and bleed, with loss of teeth; skin discolors, and wounds do not heal. Prior to the 18th century, this condition was notorious among long-duration military, particularly naval, expeditions during which participants were deprived of foods containing vitamin C.

An autoimmune disease such as lupus erythematosus or rheumatoid arthritis may attack healthy collagen fibers.

Many bacteria and viruses secrete virulence factors, such as the enzyme collagenase, which destroys collagen or interferes with its production.

Molecular Structure

A single collagen molecule, tropocollagen, is used to make up larger collagen aggregates, such as fibrils. It is approximately 300 nm long and 1.5 nm in diameter, and it is made up of three polypeptide strands (called alpha peptides, see step 2), each of which has the conformation of a left-handed helix – this should not be confused with the right-handed alpha helix. These three left-handed helices are twisted together into a right-handed triple helix or "super helix", a cooperative quaternary structure stabilized by many hydrogen bonds. With type I collagen and possibly all fibrillar collagens, if not all collagens, each triple-helix associates into a right-handed super-super-coil referred to as the collagen microfibril. Each microfibril is interdigitated with its neighboring microfibrils to a degree that might suggest they are individually unstable, although within collagen fibrils, they are so well ordered as to be crystalline.

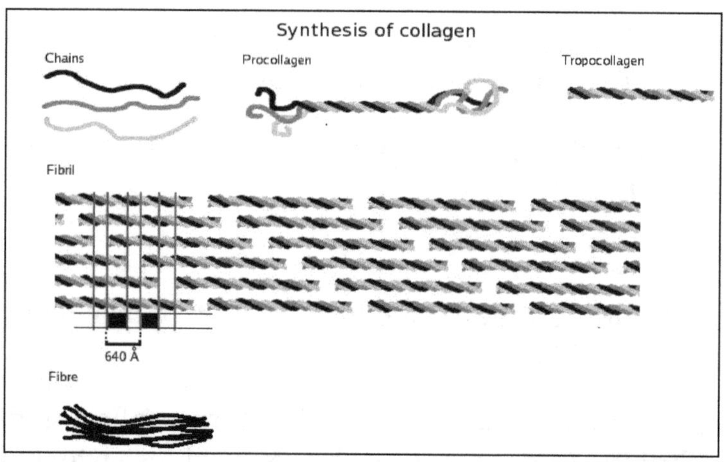

Three polypeptides coil to form tropocollagen. Many tropocollagens then bind together to form a fibril, and many of these then form a fibre.

A distinctive feature of collagen is the regular arrangement of amino acids in each of the three chains of these collagen subunits. The sequence often follows the pattern Gly-Pro-X or Gly-X-Hyp, where X may be any of various other amino acid residues. Proline or hydroxyproline constitute about 1/6 of the total sequence. With glycine accounting for the 1/3 of the sequence, this means approximately half of the collagen sequence is not glycine, proline or hydroxyproline, a fact often missed due to the distraction of the unusual GX_1X_2 character of collagen alpha-peptides. The high glycine content of collagen is important with respect to stabilization of the collagen helix as this allows the very close association of the collagen fibers within the molecule, facilitating hydrogen bonding and the formation of intermolecular cross-links. This kind of regular repetition and high glycine content is found in only a few other fibrous proteins, such as silkfibroin.

Collagen is not only a structural protein. Due to its key role in the determination of cell phenotype, cell adhesion, tissue regulation and infrastructure, many sections of its non-proline-rich regions have cell or matrix association / regulation roles. The relatively high content of proline and hydroxyproline rings, with their geometrically constrained carboxyl and (secondary) amino groups, along with the rich abundance of glycine, accounts for the tendency of the individual polypeptide strands to form left-handed helices spontaneously, without any intrachain hydrogen bonding.

Because glycine is the smallest amino acid with no side chain, it plays a unique role in fibrous structural proteins. In collagen, Gly is required at every third position because the assembly of the triple helix puts this residue at the interior (axis) of the helix, where there is no space for a larger side group than glycine's single hydrogen atom. For the same reason, the rings of the Pro and Hyp must point outward. These two amino acids help stabilize the triple helix—Hyp even more so than Pro; a lower concentration of them is required in animals such as fish, whose body temperatures are lower than most warm-blooded animals. Lower proline and hydroxyproline contents are characteristic of cold-water, but not warm-water fish; the latter tend to have similar proline and hydroxyproline contents to mammals. The lower proline and hydroxproline contents of cold-water fish and other poikilotherm animals leads to their collagen having a lower thermal stability than mammalian collagen. This lower thermal stability means that gelatin derived from fish collagen is not suitable for many food and industrial applications.

The tropocollagen subunits spontaneously self-assemble, with regularly staggered ends, into even larger arrays in the extracellular spaces of tissues. Additional assembly of fibrils is guided by fibroblasts, which deposit fully formed fibrils from fibripositors. In the fibrillar collagens, the molecules are staggered from each other by about 67 nm (a unit that is referred to as 'D' and changes depending upon the hydration state of the aggregate). Each D-period contains four plus a fraction collagen molecules, because 300 nm divided by 67 nm does not give an integer (the length of the collagen molecule divided by the stagger distance D). Therefore, in each D-period repeat of the microfibril, there is a part containing five molecules in cross-section, called the "overlap", and a part containing only four molecules, called the "gap". The triple-helices are also arranged in a hexagonal or quasihexagonal array in cross-section, in both the gap and overlap regions.

There is some covalent crosslinking within the triple helices, and a variable amount of covalent crosslinking between tropocollagen helices forming well organized aggregates (such as fibrils). Larger fibrillar bundles are formed with the aid of several different classes of proteins (including different collagen types), glycoproteins and proteoglycans to form the different types of mature tissues from alternate combinations of the same key players. Collagen's insolubility was a barrier to the study of monomeric collagen until it was found that tropocollagen from young animals can be extracted because it is not yet fully crosslinked. However, advances in microscopy techniques (i.e. electron microscopy (EM) and atomic force microscopy (AFM)) and X-ray diffraction have enabled researchers to obtain increasingly detailed images of collagen structure *in situ*. These later advances are particularly important to better understanding the way in which collagen structure affects cell–cell and cell–matrix communication, and how tissues are constructed in growth and repair, and changed in development and disease. For example, using AFM–based nanoindentation it has been shown that a single collagen fibril is a heterogeneous material along its axial direction with significantly different mechanical properties in its gap and overlap regions, correlating with its different molecular organizations in these two regions.

Collagen fibrils/aggregates are arranged in different combinations and concentrations in various tissues to provide varying tissue properties. In bone, entire collagen triple helices lie in a parallel, staggered array. 40 nm gaps between the ends of the tropocollagen subunits (approximately equal to the gap region) probably serve as nucleation sites for the deposition of long, hard, fine crystals of the mineral component, which is (approximately) $Ca_{10}(OH)_2(PO_4)_6$. Type I collagen gives bone its tensile strength.

Associated Disorders

Collagen-related diseases most commonly arise from genetic defects or nutritional deficiencies that affect the biosynthesis, assembly, postranslational modification, secretion, or other processes involved in normal collagen production.

Genetic Defects of Collagen Genes			
Type	**Notes**	**Gene(s)**	**Disorders**
I	This is the most abundant collagen of the human body. It is present in scar tissue, the end product when tissue heals by repair. It is found in tendons, skin, artery walls, cornea, the endomysium surrounding muscle fibers, fibrocartilage, and the organic part of bones and teeth.	COL1A1, COL1A2	Osteogenesis imperfecta, Ehlers–Danlos syndrome, Infantile cortical hyperostosis a.k.a. Caffey's disease

II	Hyaline cartilage, makes up 50% of all cartilage protein. Vitreous humour of the eye.	COL2A1	Collagenopathy, types II and XI
III	This is the collagen of granulation tissue, and is produced quickly by young fibroblasts before the tougher type I collagen is synthesized. Reticular fiber. Also found in artery walls, skin, intestines and the uterus	COL3A1	Ehlers–Danlos syndrome, Dupuytren's contracture
IV	Basal lamina; eye lens. Also serves as part of the filtration system in capillaries and the glomeruli of nephron in the kidney.	COL4A1, COL4A2, COL4A3, COL4A4, COL4A5, COL4A6	Alport syndrome, Goodpasture's syndrome
V	Most interstitial tissue, assoc. with type I, associated with placenta	COL5A1, COL5A2, COL5A3	Ehlers–Danlos syndrome (Classical)
VI	Most interstitial tissue, assoc. with type I	COL6A1, COL6A2, COL6A3, COL6A5	Ulrich myopathy, Bethlem myopathy, Atopic dermatitis
VII	Forms anchoring fibrils in dermoepidermal junctions	COL7A1	Epidermolysis bullosa dystrophica
VIII	Some endothelial cells	COL8A1, COL8A2	Posterior polymorphous corneal dystrophy 2
IX	FACIT collagen, cartilage, assoc. with type II and XI fibrils	COL9A1, COL9A2, COL9A3	EDM2 and EDM3
X	Hypertrophic and mineralizing cartilage	COL10A1	Schmid metaphyseal dysplasia
XI	Cartilage	COL11A1, COL11A2	Collagenopathy, types II and XI
XII	FACIT collagen, interacts with type I containing fibrils, decorin and glycosaminoglycans	COL12A1	–
XIII	Transmembrane collagen, interacts with integrin a1b1, fibronectin and components of basement membranes like nidogen and perlecan.	COL13A1	–
XIV	FACIT collagen, also known as undulin	COL14A1	–
XV	–	COL15A1	–
XVI	–	COL16A1	–
XVII	Transmembrane collagen, also known as BP180, a 180 kDa protein	COL17A1	Bullous pemphigoid and certain forms of junctional epidermolysis bullosa
XVIII	Source of endostatin	COL18A1	–
XIX	FACIT collagen	COL19A1	–
XX	–	COL20A1	–
XXI	FACIT collagen	COL21A1	–
XXII	–	COL22A1	–
XXIII	MACIT collagen	COL23A1	–
XXIV	–	COL24A1	–
XXV	–	COL25A1	–
XXVI	–	EMID2	–

XXVII	–	COL27A1	–
XXVIII	–	COL28A1	–

In addition to the above-mentioned disorders, excessive deposition of collagen occurs in scleroderma.

Diseases

One thousand mutations have been identified in 12 out of more than 20 types of collagen. These mutations can lead to various diseases at the tissue level.

Osteogenesis imperfecta – Caused by a mutation in type 1 collagen, dominant autosomal disorder, results in weak bones and irregular connective tissue, some cases can be mild while others can be lethal, mild cases have lowered levels of collagen type 1 while severe cases have structural defects in collagen.

Chondrodysplasias – Skeletal disorder believed to be caused by a mutation in type 2 collagen, further research is being conducted to confirm this.

Ehlers-Danlos syndrome – Six different types of this disorder, which lead to deformities in connective tissue, are known. Some types can be lethal, leading to the rupture of arteries. Each syndrome is caused by a different mutation, for example type four of this disorder is caused by a mutation in collagen type 3.

Alport syndrome – Can be passed on genetically, usually as X-linked dominant, but also as both an autosomal dominant and autosomal recessive disorder, sufferers have problems with their kidneys and eyes, loss of hearing can also develop in during the childhood or adolescent years.

Osteoporosis – Not inherited genetically, brought on with age, associated with reduced levels of collagen in the skin and bones, growth hormone injections are being researched as a possible treatment to counteract any loss of collagen.

Knobloch syndrome – Caused by a mutation in the COL18A1 gene that codes for the production of collagen XVIII. Patients present with protrusion of the brain tissue and degeneration of the retina, an individual who has family members suffering from the disorder are at an increased risk of developing it themselves as there is a hereditary link.

Characteristics

Collagen is one of the long, fibrous structural proteins whose functions are quite different from those of globular proteins, such as enzymes. Tough bundles of collagen called *collagen fibers* are a major component of the extracellular matrix that supports most tissues and gives cells structure from the outside, but collagen is also found inside certain cells. Collagen has great tensile strength, and is the main component of fascia, cartilage, ligaments, tendons, bone and skin. Along with elastin and soft keratin, it is responsible for skin strength and elasticity, and its degradation leads to wrinkles that accompany aging. It strengthens blood vessels and plays a role in tissue development. It is present in the cornea and lens of the eye in crystalline form. It may be one of the most abundant proteins in the fossil record, given that it appears to fossilize frequently, even in bones from the Mesozoic and Paleozoic.

Uses

Collagen has a wide variety of applications, from food to medical. For instance, it is used in cosmetic surgery and burn surgery. It is widely used in the form of collagen casings for sausages, which are also used in the manufacture of musical strings.

If collagen is subject to sufficient denaturation, e.g. by heating, the three tropocollagen strands separate partially or completely into globular domains, containing a different secondary structure to the normal collagen polyproline II (PPII), e.g. random coils. This process describes the formation of gelatin, which is used in many foods, including flavored gelatin desserts. Besides food, gelatin has been used in pharmaceutical, cosmetic, and photography industries.

Collagen adhesive was used by Egyptians about 4,000 years ago, and Native Americans used it in bows about 1,500 years ago. The oldest glue in the world, carbon-dated as more than 8,000 years old, was found to be collagen—used as a protective lining on rope baskets and embroideredfabrics, and to hold utensils together; also in crisscross decorations on human skulls. Collagen normally converts to gelatin, but survived due to dry conditions. Animal glues are thermoplastic, softening again upon reheating, so they are still used in making musical instruments such as fine violins and guitars, which may have to be reopened for repairs—an application incompatible with tough, syntheticplastic adhesives, which are permanent. Animal sinews and skins, including leather, have been used to make useful articles for millennia.

Gelatin-resorcinol-formaldehyde glue (and with formaldehyde replaced by less-toxic pentanedial and ethanedial) has been used to repair experimental incisions in rabbitlungs.

History

The molecular and packing structures of collagen have eluded scientists over decades of research. The first evidence that it possesses a regular structure at the molecular level was presented in the mid-1930s. Since that time, many prominent scholars, including Nobel laureates Crick, Pauling, Rich and Yonath, and others, including Brodsky, Berman, and Ramachandran, concentrated on the conformation of the collagen monomer. Several competing models, although correctly dealing with the conformation of each individual peptide chain, gave way to the triple-helical "Madras" model of Ramachandran, which provided an essentially correct model of the molecule's quaternary structure although this model still required some refinement. The packing structure of collagen has not been defined to the same degree outside of the fibrillar collagen types, although it has been long known to be hexagonal or quasi-hexagonal. As with its monomeric structure, several conflicting models alleged that either the packing arrangement of collagen molecules is 'sheet-like' or microfibrillar. The microfibrillar structure of collagen fibrils in tendon, cornea and cartilage has been directly imaged by electron microscopy. The microfibrillar structure of tail tendon, as described by Fraser, Miller, and Wess (amongst others), was modeled as being closest to the observed structure, although it oversimplified the topological progression of neighboring collagen molecules, and hence did not predict the correct conformation of the discontinuous D-periodic pentameric arrangement termed simply: the microfibril. Various cross linking agents like L-Dopaquinone, embeline, potassium embelate and 5-O-methyl embelin could be developed as potential cross-linking/stabilization agents of collagen preparation and its application as wound dressing sheet in clinical applications is enhanced.

References

- Jones, Daniel (2003), Peter Roach, James Hartmann and Jane Setter, eds., English Pronouncing Dictionary, Cambridge: Cambridge University Press, ISBN 3-12-539683-2
- Duggleby RG. "Analysis of enzyme progress curves by non-linear regression". Methods in Enzymology. Methods in Enzymology. 249: 61–90. doi:10.1016/0076-6879(95)49031-0. ISBN 978-0-12-182150-0. PMID 7791628
- Fersht, Alan. Structure and mechanism in protein science: a guide to enzyme catalysis and protein folding. San Francisco: W.H. Freeman. ISBN 0-7167-3268-8
- Danson, Michael; Eisenthal, Robert (2002). Enzyme assays: a practical approach. Oxford [Oxfordshire]: Oxford University Press. ISBN 0-19-963820-9
- Walsh, Ryan (2012). "Ch. 17. Alternative Perspectives of Enzyme Kinetic Modeling". In Ekinci, Deniz. Medicinal Chemistry and Drug Design(PDF). InTech. pp. 357–371. ISBN 978-953-51-0513-8
- Gibson QH. "Rapid mixing: Stopped flow". Methods in Enzymology. Methods in Enzymology. 16: 187–228. doi:10.1016/S0076-6879(69)16009-7. ISBN 978-0-12-181873-9
- Maton, Anthea; Jean Hopkins; Charles William McLaughlin; Susan Johnson; Maryanna Quon Warner; David LaHart; Jill D. Wright (1993). Human Biology and Health. Englewood Cliffs, New Jersey, USA: Prentice Hall. ISBN 0-13-981176-1
- Steinberg, MH (2001). Disorders of Hemoglobin: Genetics, Pathophysiology, and Clinical Management. Cambridge University Press. p. 95. ISBN 0-521-63266-8
- Nelson DL, Cox MM (2000). Lehninger Principles of Biochemistry (3rd ed.). New York: Worth Publishers. p. 206. ISBN 0-7167-6203-X

- Hall, John E. (2010). Guyton and Hall textbook of medical physiology (12th ed.). Philadelphia, Pa.: Saunders/Elsevier. p. 502. ISBN 978-1416045748
- William, C. Wilson; Grande, Christopher M.; Hoyt, David B. (2007). "Pathophysiology of acute respiratory failure". Trauma, Volume II: Critical Care. Taylor & Francis. p. 430. ISBN 978-1-4200-1684-0
- Guyton, Arthur C.; John E. Hall (2006). Textbook of Medical Physiology (11 ed.). Philadelphia: Elsevier Saunders. p. 511. ISBN 0-7216-0240-1
- Handin, Robert I.; Lux, Samuel E. and StosselBlood, Thomas P. (2003). Blood: Principles & Practice of Hematology. Lippincott Williams & Wilkins, ISBN 0781719933

Permissions

All chapters in this book are published with permission under the Creative Commons Attribution Share Alike License or equivalent. Every chapter published in this book has been scrutinized by our experts. Their significance has been extensively debated. The topics covered herein carry significant information for a comprehensive understanding. They may even be implemented as practical applications or may be referred to as a beginning point for further studies.

We would like to thank the editorial team for lending their expertise to make the book truly unique. They have played a crucial role in the development of this book. Without their invaluable contributions this book wouldn't have been possible. They have made vital efforts to compile up to date information on the varied aspects of this subject to make this book a valuable addition to the collection of many professionals and students.

This book was conceptualized with the vision of imparting up-to-date and integrated information in this field. To ensure the same, a matchless editorial board was set up. Every individual on the board went through rigorous rounds of assessment to prove their worth. After which they invested a large part of their time researching and compiling the most relevant data for our readers.

The editorial board has been involved in producing this book since its inception. They have spent rigorous hours researching and exploring the diverse topics which have resulted in the successful publishing of this book. They have passed on their knowledge of decades through this book. To expedite this challenging task, the publisher supported the team at every step. A small team of assistant editors was also appointed to further simplify the editing procedure and attain best results for the readers.

Apart from the editorial board, the designing team has also invested a significant amount of their time in understanding the subject and creating the most relevant covers. They scrutinized every image to scout for the most suitable representation of the subject and create an appropriate cover for the book.

The publishing team has been an ardent support to the editorial, designing and production team. Their endless efforts to recruit the best for this project, has resulted in the accomplishment of this book. They are a veteran in the field of academics and their pool of knowledge is as vast as their experience in printing. Their expertise and guidance has proved useful at every step. Their uncompromising quality standards have made this book an exceptional effort. Their encouragement from time to time has been an inspiration for everyone.

The publisher and the editorial board hope that this book will prove to be a valuable piece of knowledge for students, practitioners and scholars across the globe.

Index

www.ingramcontent.com/pod-product-compliance
Lightning Source LLC
LaVergne TN
LVHW080704210926
848485LV00003B/418